KU-051-134

Extinctions in the History of Life

Extinction is the ultimate fate of all biological species – over 99 per
cent of the species that have ever inhabited the Earth are now extinct.
The long fossil record of life provides scientists with crucial
information about when species became extinct, which species were
most vulnerable to extinction and what processes may have brought
about extinctions in the geological past. Key aspects of extinctions in
the history of life are here reviewed by six leading palaeontologists,
providing a source text for geology and biology undergraduates as
well as more advanced scholars. Topical issues such as the causes of
mass extinctions and how animal and plant life has recovered from
these cataclysmic events that have shaped biological evolution are
dealt with. This helps us to view the current biodiversity crisis in a
broader context, and shows how large-scale extinctions have had
profound and long-lasting effects on the Earth's biosphere.

PAUL TAYLOR is former Head of Invertebrates and Plants at The
Natural History Museum, London. His research on bryozoans has been
acknowledged with the Paleontological Society's Golden Trilobite
Award (1993), and a Distinguished Scientists Award from UCLA (2002).
He has edited or coedited three books: *Major Evolutionary Radiations*
(with G. P. Larwood; 1990. Clarendon Press, Oxford), *Biology and
Palaeobiology of Bryozoans* (with P. J. Hayward and J. S. Ryland; 1995.
Olsen & Olsen, Fredensborg), and *Field Geology of the British Jurassic*
(1995. Geological Society of London). He is also the author of the
Dorling Kindersley Eyewitness book *Fossil* (1990), and has published
more than 150 scientific articles.

Central Library
Y Llyfrgell Ganolog
☎ 02920 38211

ACC. No: 02699754

Contents

Notes on contributors

David J. Bottjer

Born in New York City and educated in geology at Haverford College
(B.S. 1973), the State University of New York at Binghampton (M.A.
1976), and Indiana University (Ph.D. 1978), David J. Bottjer began his
career as a National Research Council–USGS Postdoctoral Fellow at
the National Museum of Natural History, Smithsonian Institution. In
1979 he joined the faculty of the Department of Earth Sciences at the
University of Southern California, where he is currently Professor of
Paleontology and also a Research Associate at the nearby Natural
History Museum of Los Angeles County. Editor-in-Chief of the
internationally renowned journal *Palaeogeography, Palaeoclimataology,
Palaeoecology*, and co-editor of the book series *Critical Moments and
Perspectives in Paleobiology and Earth History*, Dr Bottjer has lectured
throughout the world, most recently in Switzerland, New Zealand,
Japan and the UK. A 1992–93 Paleontological Society Distinguished
Lecturer, a Fellow both of the American Association for the
Advancement of Science and the Geological Society of America, and
President of the Pacific Coast Section of the Society for Sedimentary
Geology, in 2000 he was a Visiting Fellow at CSEOL, UCLA. Dr Bottjer's
research centres on the evolutionary palaeoecology of
macroinvertebrate animals in the Phanerozoic fossil record.

David Jablonski

Educated in geology at Columbia University (B.A. 1974) and Yale
University (M.S. 1976, Ph.D. 1979), David Jablonski became enamoured
with fossils at an early age, working as an undergraduate at New
York's American Museum of Natural History. Following postdoctoral
studies at UC Santa Barbara and UC Berkeley, he spent three years on

the biology faculty at the University of Arizona before joining the
University of Chicago in 1985 where he is currently William R. Kenan,
Jr., Professor in the Department of Geophysical Sciences and Chair of
the Committee on Evolutionary Biology. He holds a joint appointment
with the Field Museum of Natural History in Chicago, and is an
Honorary Research Fellow at The Natural History Museum in London.
A very active contributor to his profession nationally and
internationally, Dr Jablonski has also led University of Chicago alumni
tours to the Galapagos Islands, the Gulf of California,
Yucatan–Belize–Honduras–Guatemala, and Alaska–British Columbia.
Co-editor of three major scientific volumes, he is a fellow in the
American Academy of Arts and Sciences and recipient both of the
Schuchert Award of the Paleontological Society and a Guggenheim
Fellowship. Dr Jablonski's research centres on large-scale patterns in
the evolutionary history of marine invertebrate animals as revealed by
the fossil record.

J. William Schopf

Director of UCLA's Center for the Study of Evolution and the Origin of
Life (CSEOL) and a member of the Department of Earth and Space
Sciences, J. William Schopf received his undergraduate training in
geology at Oberlin College, Ohio, and in 1968 his PhD degree in
biology from Harvard University. He has edited eight volumes,
including two prize-winning monographs on early evolution – his
primary research interest – and is author of *Cradle of Life*, awarded
Phi Beta Kappa's 2000 national science book prize. At UCLA, he has
been honoured as a Distinguished Teacher, a Faculty Research
Lecturer, and as recipient of the university-wide Gold Shield Prize
for Academic Excellence. A Humboldt Fellow in Germany and a
foreign member both of the Linnean Society of London and the
Presidium of the Russian Academy of Science's A. N. Bach Institute
of Biochemistry, Dr Schopf is a member of the National Academy of
Sciences and the American Philosophical Society, a fellow of the
American Academy of Arts and Sciences, and current President of the
International Society for the Study of the Origin of Life (ISSOL). Listed
by *Los Angeles Times Magazine* as among southern California's most
outstanding scientists of the twentieth century, he is recipient of
medals awarded by ISSOL, the National Academy of Sciences, and the
National Science Board, and has twice been awarded Guggenheim
Fellowships.

Paul D. Taylor

Born in Hull, England, Paul Taylor received his undergraduate degree (BSc in Geology, 1974) from the University of Durham and stayed there to complete a PhD in 1977. After undertaking a postdoctoral fellowship under the guidance of Derek Ager at the University College of Swansea, in 1979 he joined the staff of the then British Museum (Natural History), now The Natural History Museum, as a researcher in the Department of Palaeontology. From 1990 until 2003 he served as Head of the Invertebrates and Plants Division. Dr Taylor has carried out scientific fieldwork in various parts of the world, including Saudi Arabia, India, New Zealand, Russia, Spitsbergen, several European countries and the USA. He has held Visiting Research positions at the University of Otago (New Zealand), the Museum National d'Histoire Naturelle (Paris), CSEOL (UCLA) and Hokkaido University (Japan). Fellow of the Linnean Society of London and author or editor of four books and more than 150 scientific articles, he has served on various national and international scientific committees and editorial boards, and is currently President of the International Bryozoology Association. In 1992 he was co-recipient of the Paleontological Society's award for the most outstanding monograph in systematic palaeontology. Dr Taylor's research centres on the taxonomy and palaeobiology of bryozoans, a group of colonial marine invertebrates with a rich fossil record.

Paul Wignall

A native of Bradford, England, Paul Wignall received his education in geology at Oxford University (BSc 1985) and the University of Birmingham (PhD 1988). Following a year of postdoctoral research in the laboratory of Professor John Hudson at the University of Leicester, in 1989 he joined the School of Earth Sciences at the University of Leeds where he is now Reader in Palaeoenvironments. An expert on the origin of marine petroleum and the palaeoecology of oil-source rocks, Dr Wignall's research has also focused on the causes of mass extinctions, particularly that at the end of the Permian. This interest has taken him on fieldwork to China, Pakistan, Greenland, Italy, Austria, Spitsbergen, Tibet and the USA. Author of two books (*Black Shales* and *Mass Extinctions and their Aftermath*, the latter co-authored with his doctoral supervisor Professor A. Hallam), and a member of the editorial boards of major international journals, he is a recipient of the President's Award of the Geological Society of London, the

Fearnside's Prize of the Yorkshire Geological Society and the Clough Award of the Edinburgh Geological Society.

Scott L. Wing

A palaeobotanist and palaeoecologist educated in biology at Yale University (B.S. 1976; Ph.D., 1981), Scott L. Wing began his career as a National Research Council–US Geological Survey (USGS) Postdoctoral Fellow (1982–83) and Geologist (1983–84) in the USGS Paleontology and Stratigraphy Branch. In 1984 he joined the staff of the Department of Paleobiology at the National Museum of Natural History (NMNH), Smithsonian Institution, where he has risen through the ranks to his current position of Research Curator, and since 1992 has held a joint appointment in the Department of Earth and Environmental Sciences at the University of Pennsylvania. He served for six years as Co-Editor of *Paleobiology*, the prestigious journal published by the Paleontological Society, has co-edited four major volumes, and is currently a member of the editorial boards of *Evolutionary Ecology Research* and *Annual Reviews of Ecology and Systematics*. Over the past decade, he and his colleagues at the NMNH have organized briefings for Congress and federal agencies as well as symposia at national and international conferences. Dr Wing's main areas of interest are the effects of climate change and global warming on the world's biota, especially the vegetation, as evidenced by the fossil record.

Preface

Extinction is a corollary of life itself. Just as the death of individuals is assured, so the extinction of species can be pretty much guaranteed in the fullness of geological time. Indeed, a leading palaeontologist once famously quipped that to a first approximation life on Earth is extinct. By this he meant that the great majority of species ever to have lived on the planet are no longer with us. Today we are rightly concerned with the threat to the survival of many contemporary species, and we mourn the loss of those that have disappeared in historic times, more especially because their extinction was very often due to overexploitation or habitat destruction by humankind. While the extinctions occurring at the present day may be viewed as atypical and in some respects 'unnatural', taking a broader view across geological time extinction can be seen as a major constructive force in the evolution of life, removing incumbents and allowing other groups of animals and plants to prosper and diversify. A renaissance of interest in extinction has been ignited not only by the contemporary biodiversity crisis, but also by the development of analytical approaches to the fossil record and of new geological techniques that have greatly increased our appreciation of global change. Our understanding of extinctions in the history of life is far better now than it was a few decades ago.

This publication arises from a symposium held at the University of California, Los Angeles and convened by the Center for the Study of Evolution and the Origin of Life (CSEOL). Our aim, both in the symposium and in this book, has been to make accessible – at undergraduate level – key findings and current debates concerning extinctions in the history of life. Chapter 1 introduces the topic and sets the scene for the five chapters that follow. The 'rules' of the extinction game played out during the Precambrian when most life was microbial are shown in Chapter 2 to have been different from those of later times. Continuing the non-animal theme, Chapter 3 focuses on plants and asks whether they have suffered similar mass extinctions to those that have periodically wreaked havoc among animals. Chapter 4 takes a detailed look at a prolonged interval of geological time characterized by high levels of environmental stress and sustained extinction. The various processes implicated in mass extinctions are reviewed in Chapter 5. Finally, Chapter 6 rounds off the book by considering the evolutionary role of mass extinctions. A glossary of terms has been included to assist the reader.

Gratitude is owed to various people who helped with the symposium and/or the production of this volume: Richard Mantonya, Bill Schopf, Bill Clemmens, Nicole Fraser, Paul Kenrick and Patricia Taylor. Bonnie Dalzell generously allowed reproduction of her magnificent illustration (Figure 6.2) of a gigantic extinct bird.

PAUL D. TAYLOR

Department of Palaeontology, The Natural History Museum, London, UK

1
_____

Extinction and the fossil record

INTRODUCTION

The fossil record provides us with a remarkable chronicle of life on Earth. Fossils show how the history of life is characterized by unending change – species originate and become extinct, and clades wax and wane in diversity through the vastness of geological time. One thing is clear – extinction has been just as important as the origination of new species in shaping life's history.

It has been estimated that more than 99 per cent of all species that have ever lived on Earth are now extinct. While species of some prokaryotes may be extremely long-lived (Chapter 2), species of multicellular eukaryotes in the Phanerozoic fossil record commonly become extinct within 10 million years (Ma) of their time of origin, with some surviving for less than a million years. Entire groups of previously dominant animals and plants have succumbed to extinction, epitomized by those stalwarts of popular palaeontology, the dinosaurs. The extinction of dominant clades has had positive as well as negative consequences – extinction removes incumbents and opens the way for other clades to radiate. For example, without the extinction of the incumbent dinosaurs and other 'ruling reptiles' 65 Ma ago, birds and mammals, including humans, would surely not be the dominant terrestrial animals they are today.

Over the past 30 years palaeontologists have increasingly turned their attention towards the documentation of evolutionary patterns and the interpretation of processes responsible for these patterns. As part of this endeavour, extinction has become a major focus of study. Mass extinctions – geological short intervals of time when the Earth's

Extinctions in the History of Life, ed. Paul D. Taylor.
Published by Cambridge University Press. © Cambridge University Press 2004.

biota was very severely depleted – have received particular attention for two main reasons. First, new geological evidence has been obtained for the causes of mass extinctions (Chapter 5). Second, it has become apparent that the sudden and catastrophic events precipitating mass extinctions have the potential to exterminate species with scant regard for how well they were adapted to normal environmental conditions; the rules of the survival game may change drastically during these times of global catastrophe (Chapter 6). Uniformitarianism, explaining geological phenomena through the action of the slow and gradual processes we can observe in daily operation, has been the guiding paradigm for geologists since Charles Lyell (1797–1875) published *Principles of Geology* in 1830. However, uniformitarianism alone is insufficient to explain how extinction has moulded the history of life; catastrophic events have also played a key role and this realization is reflected by a revival of scientific interest in catastrophism.

Studies of extinctions in the geological past are relevant in providing a broader context, potentially with remedial lessons, for the contemporary biodiversity crisis being driven largely by the activities of humankind. Published data indicate an accelerating rate of extinction of mammal and bird species for each 50-year interval since 1650. Between 60 and 88 mammal species are thought to have become extinct during the last 500 years, representing about two per cent of the total diversity. Perhaps the most notorious of these extinctions occurred in the late seventeenth century with the disappearance of the Dodo (*Raphus cucullatus*), a large flightless pigeon from the island of Mauritius in the Indian Ocean, immortalized (in words only) by the phrase 'as dead as a Dodo'. Between 11 and 13 per cent of bird and plant species living today are thought to be close to extinction. A pessimistic estimate considers that up to 50 per cent of the world's biota could face extinction within the next 100 years. Current rates of extinction for relatively well-known groups may be 100 to 1000 times greater than they were during pre-human times (Pimm *et al.*, 1995). Concerns about the human threat to contemporary biodiversity are equally valid for organisms living in the sea as they are for the better known terrestrial biota – the coastal marine environment has been severely disturbed and depleted of diversity by overfishing (Jackson, 2001). While there is no evidence that major extinctions in the geological past resulted from comparable over-exploitation by a single species, there are general lessons to be learnt from ancient extinctions. The most sobering of these lessons is that the Earth's biota recovers extremely slowly after major extinction events. Ten million years or more may elapse before biotas have returned to something like their previous levels of diversity.

This chapter aims to give an introduction to extinctions in the fossil record, setting the scene and providing a general background for the more detailed accounts in the chapters that follow. After a brief historical preface, I describe how extinction is detected and measured in the fossil record, the broad patterns of extinction and biodiversity change that are evident in the Phanerozoic fossil record, and the interpretation of extinction patterns and processes.

BRIEF HISTORY OF FOSSIL EXTINCTION STUDIES

Two hundred years ago there was no general agreement among naturalists that any species had ever become extinct. Although naturalists at that time knew of fossil species that had never been observed alive, most of these were marine animals and it remained possible that they would eventually be discovered living somewhere in the poorly explored seas and oceans of the world. The great French naturalist Georges Cuvier (1769–1832; Figure 1.1A) is generally accredited with establishing the reality of ancient extinctions. Cuvier's work on the fossils from Cenozoic deposits in and around Paris revealed the former existence of several species of large terrestrial mammals ('quadrapeds') not known to be living at the present day but which would surely have been discovered if they had been: 'Since the number of quadrapeds is limited, and most of their species – at least the large ones – are known, there are greater means to check whether fossil bones belong to one of them, or whether they come from a lost species.' (Cuvier, 1812, translation in Rudwick, 1997). Cuvier promoted the idea of catastrophism to explain the extinction of species. According to him, major geological upheavals, unlike anything witnessed by humankind during modern times, were responsible for these extinctions. One of Cuvier's main reasons for favouring catastrophic extinction was his belief that species were so well-adapted that their gradual extinction was inconceivable (Rudwick, 1997).

Alcide d'Orbigny (1802–57; Figure 1.1B), a student of Cuvier's who undertook detailed research on the taxonomy and stratigraphical distribution of fossil invertebrates, extended his mentor's ideas. D'Orbigny's findings led him to propose that life on Earth had been devastated by 27 catastrophic extinctions. All living species were exterminated during each extinction event, subsequently to be replaced by a totally new biota formed in a fresh creation of life. The stratigraphical stages (e.g. Bajocian, Cenomanian) erected by d'Orbigny, still used by geologists

Figure 1.1. Portraits of some pioneers in the early study of fossil extinctions. A, Georges Cuvier; B, Alcide d'Orbigny; C, Charles Darwin; D, John Phillips.

today in a modified way, each represent an interval of geological time when the Earth was populated by one of the 27 biotas. In contrast to the catastrophist creationists Cuvier and d'Orbigny, Charles Darwin (1809–82; Figure 1.1C) was an evolutionist who followed the uniformitarian principles expounded by Lyell. He believed in the gradual disappearance of species, one after the other, rather than their sudden decimation. He considered that natural selection was sufficient to explain the extinction of species, writing in the *Origin of Species* (1859) that 'the improved and modified descendants of a species will generally cause the extermination of the parent-species.'

The British geologist John Phillips (1800–74; Figure 1.1D) made an early attempt to estimate the broad changes in the diversity of life on Earth between the Cambrian and the present day. Phillips' (1860) plot of diversity against time shows two major drops, one at the end of the Palaeozoic and the second at the end of the Mesozoic. Through his first-hand experience of fossils and their distribution in strata of different ages, Phillips was able to recognize the great turnovers of life that marked the transitions between the Palaeozoic, Mesozoic and Cenozoic eras. The wholescale extinctions of species at these era bound-aries we now call the end-Permian and Cretaceous–Tertiary (K–T) mass extinctions (Figure 1.2).

Skipping forward to the final quarter of the twentieth century, the contributions of two research groups ignited a major resurgence of interest in extinctions in the fossil record. Jack Sepkoski's compi-lation of the ranges through the Phanerozoic of marine families, and later of marine genera, opened the way for the analysis of global extinc-tion patterns undertaken in collaboration with David Raup. A landmark paper (Raup and Sepkoski, 1982) on extinction rates enabled five mass extinctions to be recognized in the Phanerozoic, and a subsequent anal-ysis (Raup and Sepkoski, 1984) suggested a 26-Ma periodicity in extinc-tion between the end of the Permian and the present day. The first of these papers stimulated a flurry of work among palaeontologists inter-ested in how particular taxonomic groups had fared during these 'Big Five' mass extinctions (e.g. Larwood, 1988), while the claim of periodic-ity prompted various astronomical explanations of causal mechanisms, engagingly summarized by Raup (1986).

At about the same time that Sepkoski and Raup were compil-ing and analysing data from the fossil record, a team led by Luis and Walter Alvarez at UC Berkeley discovered an enrichment of the element iridium at the K–T boundary in Gubbio, Italy (Alvarez *et al.*, 1980), coincident with the K–T mass extinction which removed the last dinosaurs and many other species. This iridium anomaly, later iden-tified at the same stratigraphical level elsewhere in the world, pro-vided strong evidence for the impact of a sizeable extraterrestrial object (bolide or asteroid) with the Earth, an impact with numerous possible consequences devastating to life on the planet. Initially received with scepticism by most palaeontologists, the impact hypothesis for the K–T mass extinction has since won considerable support. Identification of the apparent impact crater (Chicxulub, Mexico) and a wealth of other

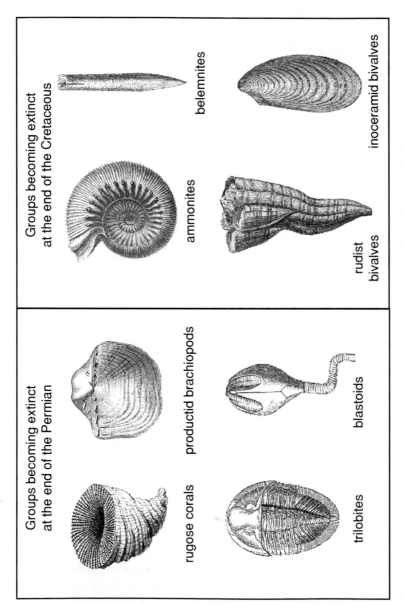

Figure 1.2. Examples of species belonging to marine invertebrate groups that suffered extinction during the end-Permian and end-Cretaceous (K–T) mass extinctions. Note that the species depicted were *not* among the final survivors of their groups, and that rudists as a whole may have succumbed a little before the K–T event. Original lithographs taken from Nicholson (1879).

geological evidence (shocked quartz grains, tektites, tsunami deposits, etc.) have corroborated the original hypothesis, although the kill mechanism/s and the possible involvement of other environmental changes in the K–T mass extinction are still contentious issues (Chapter 5).

DETECTING AND MEASURING EXTINCTIONS

What exactly is extinction?

Extinction is quite simply the 'death' of a taxon. The extinction of a species occurs when the last individual of that species dies. The extinction of a genus happens when the last individual belonging to the last species of the genus dies, and so on. In the case of a small number of species that have become extinct in historical times, the death of the last individual, and therefore the extinction of the species, has actually been observed. For example, the last Tasmanian tiger (*Thylacinus cyanocephalus*), a wolf-like marsupial mammal, died in Hobart Zoo on 7 September 1936. Nonetheless, unsubstantiated sightings of Tasmanian tigers in the wild are still occasionally reported, illustrating that even for such a large and distinctive contemporary animal it can be difficult to verify extinction.

Detecting extinction in the incomplete fossil record

Pinpointing the moment of extinction is much more of a problem in the fossil record than it is among organisms that became extinct during historical times. Even large-scale extinctions seldom generate mass mortality deposits (cf. Zinsmeister, 1998) where palaeontologists might expect or hope to find the last individuals belonging to a species. No palaeontologist would ever claim that a particular fossil specimen represents the very last survivor of a species – the probability of this last individual being fossilized, discovered and collected are infinitesimally small. Even if we did have this fossil to hand we could never be sure that this is what it was. A fundamental difficulty with extinction is that it is impossible to prove a negative – the absence of a species – and therefore to be sure exactly when extinction occurred. Nevertheless, repeated interrogation of the fossil record does allow scientists to corroborate and refine assessments of when a species, or a clade, became extinct. We can, for instance, be confident that the last ammonites became extinct at or before the K–T boundary because intensive sampling of younger, post-Cretaceous rocks has failed to produce any unequivocally indigenous ammonites. The last appearance of a species (or genus, family, etc.) in the fossil record seldom coincides with the time of its extinction. Instead, the last appearance will always precede the true time of extinction because of the incompleteness of the fossil record. Stratigraphical completeness sets an upper limit on the completeness of the fossil record – the fossil record can never be more complete than the rocks containing the fossils. Schindel (1982) calculated completeness for seven stratigraphical successions which had been used in evolutionary studies because they were regarded as being relatively complete. Even in these exceptionally complete successions, stratigraphical completeness never exceeded 45 per cent and was 10 per cent or less for five of the seven successions.

Signor and Lipps (1982) realized that sampling gaps in the fossil record could make the severity of a mass extinction event seem less than it actually was. The last appearances of taxa before their true time of extinction are 'smeared' through an interval of time before the mass extinction. This sampling artefact is termed a Signor–Lipps Effect (Figure 1.3). For example, Rampino and Adler (1998) showed how a Signor–Lipps Effect could account for the extinction pattern of foraminifera before the end-Permian mass extinction in the Italian Alps. Species of foraminifera with overall lower abundances through a sequence of Permian rocks are the first to disappear from the fossil record, whereas more abundant species range higher in the section, as would be predicted if their last appearances in the fossil record were determined by sampling.

Another pattern resulting from the incompleteness of the fossil record occurs when a species apparently becomes extinct only to reappear in younger rocks (Figure 1.3). This is known as a Lazarus Effect, after the disciple who reputedly returned from the dead. Taxa missing from the fossil record but which can be inferred to have been alive at the time by their occurrence in both older and younger rocks are called Lazarus taxa (see Fara, 2001). Lazarus taxa are useful in assessing the quality of the fossil record – the greater the proportion of Lazarus taxa present during a given interval of geological time, the poorer is the fossil record for that time interval. Times of apparently high extinction intensity may sometimes be reinterpreted as due to deficiencies in the fossil record when a high proportion of Lazarus taxa are present. However, Lazarus taxa do often increase in number at times of true mass extinction (e.g. Twitchett, 2001) because the same factors that bring about the genuine extinction of some taxa may cause other taxa to shrink in geographical range and/or population size (Wignall

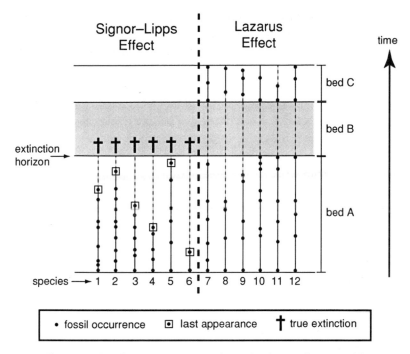

Figure 1.3. Two important patterns for extinction studies caused by gaps in the fossil record illustrated by range data for 12 hypothetical species with respect to an extinction horizon. The backward smearing of last appearances of species that became extinct at a major extinction event is known as a Signor–Lipps Effect. Temporary disappearance of taxa through an interval of time, often associated with true extinction of other taxa, produces a Lazarus Effect.

and Benton, 1999), removing them temporarily from the fossil record until more normal conditions return. Survival of Lazarus taxa is usually explained by postulating the existence of refuges – safe places where the adverse factors causing extinction were absent or reduced. For every extinction event, there are always particular habitats and/or geographical regions not represented in the fossil record that might have provided refuges for Lazarus taxa.

Pseudoextinctions in the fossil record should be distinguished from true extinctions. The subdivision of an evolving lineage into two or more species results in the pseudoextinction of the ancestral species at each transition. It would be incorrect to classify this as a true extinction because genetic continuity is maintained between ancestral and descendant species – the branch on the evolutionary tree is not terminated but continues under a different name. The change of species name is sometimes placed at an arbitrary point within the lineage, for example coinciding with a geological boundary, or a time of particularly rapid morphological change. Alternatively, it may be made at an abrupt morphological jump.

A second kind of pseudoextinction can occur when we deal with taxa above the species level. The pseudoextinction of paraphyletic higher taxa (paraclades) is best illustrated using a frequently cited example, the dinosaurs. Dinosaurs of popular understanding – large, landdwelling animals such as *Brachiosaurus* and *Tyrannosaurus* – are a paraclade. Only when birds are included within dinosaurs do we get a true clade. This is because birds are more closely related to some dinosaurs than these dinosaurs are to other dinosaurs, that is birds and these dinosaurs share a more recent common ancestor. While there is little doubt that the last dinosaur species of popular understanding did become extinct during the K–T mass extinction, birds survived this extinction, carrying their dinosaurian genetic legacy through to the present day. Leaving aside semantic aspects, the importance for extinction pattern analysis of including or excluding pseudoextinctions of paraclades in databases has been debated vigourously by palaeontologists. Until a great many more phylogenetic analyses have been completed, most fossil data on extinction patterns will inevitably comprise a mixture of true clades and paraclades. Simulation studies suggest, however, that inclusion of paraclades does not substantially alter extinction patterns, and even the loss of a paraclade, such as the dinosaurs, involves the true extinction of one or more species.

Measuring extinction

A variety of metrics have been applied to quantify extinction in the fossil record (Figure 1.4). The simplest is number of extinctions (E), i.e. the number of taxa becoming extinct. This measure has limited applicability in studies of extinctions through geological time because it is dependent on the duration of the time interval in question and on the number of taxa present during that interval. Geological time is not conventionally divided into slices of even duration – stratigraphical stages, a commonly used division, differ substantially in their durations. A large value of E may simply reflect a time interval of greater than average length. Therefore, extinction rate (E/t) is often calculated, usually expressed per million years. Another metric – per taxon extinction (E/D) – allows for differences in diversity by dividing the number of extinctions by the diversity (D) of taxa present during the time interval

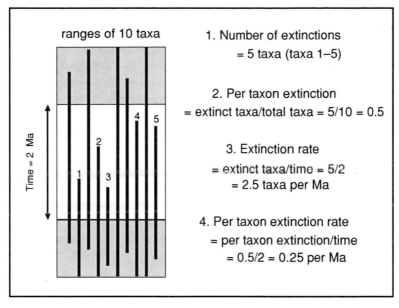

Figure 1.4. Metrics used to quantify extinction intensity. Range bars for 10 taxa are shown, all present within the shaded stratigraphical interval of interest which has been assigned an arbitrary duration of 2 million years (Ma). Five of the taxa have their last appearances within this interval, i.e. suffer extinction. This is equivalent to a per taxon extinction of 0.5 (or 50 per cent), a rate of extinction of 2.5 taxa per Ma, and a per taxon extinction rate of 0.25 per Ma.

in question. A third commonly used metric is per taxon extinction rate ($E/D/t$). While compensating for variations in both duration and diversity has clear advantages, errors can be introduced if there are uncertainties in the length of time represented by the interval or in the actual diversity of taxa present. The use of different extinction metrics can sometimes affect perceptions of extinction patterns. For example, the relative severities of bryozoan genus extinction in the terminal stage of the Cretaceous (Maastrichtian) and basal stage of the Tertiary (Danian) change according to the extinction metric used (McKinney and Taylor, 2001).

Extinction studies vary in their taxonomic scope and level in the taxonomic hierarchy (e.g. species, genus, family), stratigraphical precision and geographical coverage. Constraints imposed by the imperfect fossil record, discussed above, mean that we will never know the exact time of global extinction for all species belonging to all taxonomic groups. Instead, we must make interpretations about extinctions from data that is more restricted in taxonomic and geographical scope and less precise in time. Because palaeontologists tend to specialize on particular taxonomic groups, much of the extinction literature deals with single taxonomic groups such as phyla. Not surprisingly, the extinction pattern evident in one taxonomic group may not match that seen in another group, particularly if the groups inhabited very different environments (e.g. land or sea). Palaeontologists studying different taxonomic groups may therefore develop contrasting views of the importance of extinction events.

Local and regional studies of extinction are sometimes undertaken at species level but relatively few global studies have dealt with species, usually because of the difficulty of mining data from the voluminous palaeontological literature. As an alternative, genera or families are commonly used. The rationale here is that the extinction of genera or families provides a reasonable proxy for species extinction, i.e. extinction patterns above species level parallel those at species level. While this is indeed often the case, situations can exist where extinction rate changes at species level are not matched proportionally at genus or family level, when the extinct species are not uniformly distributed among genera and families.

Much of the research on global extinction patterns has been undertaken at the level of the stratigraphical stage. Stages vary in duration but typically comprise time slices of between 4 and 15 Ma. The stage is frequently the finest division of geological time that can be achieved when mining data on taxonomic ranges from the literature. In much of the early palaeontological literature and in many museum collections the exact stratigraphical horizon from which fossils were collected is not stated and even stage-level precision is impossible to achieve. Bed-by-bed sampling is a favoured approach of palaeontologists studying extinctions through local stratigraphical sequences. When combined with modern techniques of stratigraphical correlation, including graphic correlation, this permits a significantly greater degree of stratigraphical subdivision than the stage. Considerable resampling of the fossil record will be needed in the future if we are to refine the stratigraphical precision of data on extinction patterns.

Finally, our knowledge of the fossil record is far from equal for all parts of the world. North America and Europe have better documented fossil records than elsewhere (Smith, 2001). Accordingly, these regions contribute disproportionately to so-called 'global' extinction patterns. While improved knowledge of the fossil record from other parts of the world is unlikely to overturn the major extinction patterns apparent in the known fossil record, it may force re-evaluation of some of the details, including the times of extinction for particular taxa and the significance of minor mass extinction events.

The Phanerozoic encompasses nearly all of the fossil record of macrofossils, i.e. large and complex multicellular organisms. This eon of geological time began about 545 Ma ago at the dawn of the Cambrian period, and continues through to the present day. During the past 30 years, Phanerozoic diversity patterns have been intensively studied in both the marine and terrestrial realms, although it is the former that has provided the best data on extinction patterns. Sepkoski's pioneering compilations of the ranges of marine familes and genera permitted the first detailed descriptions and analyses of Phanerozoic diversity, origination and extinction patterns (Sepkoski, 1981). The general findings of his work have been supported by an independent family-level compilation – *The Fossil Record 2* (Benton, 1993) – that also includes terrestrial animals and plants.

Marine invertebrate families show a characteristic pattern of diversity change through the Phanerozoic (Figure 1.5). The dominant features of this pattern are: (1) a steep increase in diversity between the basal Cambrian and Ordovician; (2) a Palaeozoic diversity plateau from Ordovician to Permian; (3) a rapid decline in diversity at the Permian–Triassic boundary (i.e. Palaeozoic–Mesozoic transition); and (4) a sustained increase in diversity between the Triassic and the present day to a level approximately twice that of the Palaeozoic plateau. Minor fluctuations in diversity are superimposed over the general pattern, producing subsidiary peaks and troughs. At genus level the Palaeozoic plateau becomes a gentle incline, with generic diversity decreasing steadily from a Late Ordovician high to the Permian (Figure 1.6). The generic diversity curve is also more jagged than the family curve.

Diversity plots for insect and non-marine tetrapod families are also shown in Figure 1.5. The fossil record of both of these groups began later than marine invertebrates, in the Devonian. Insects exhibit a sustained rise in diversity, with minor fluctuations among which the drop in diversity at the Permian–Triassic boundary stands out. There is no indication of significant levels of insect extinction at the end of the Cretaceous. The relatively low family diversity of non-marine tetrapods, especially prior to 100 Ma ago when modern groups began to diversify, makes it difficult to see details of their diversity history at the scale

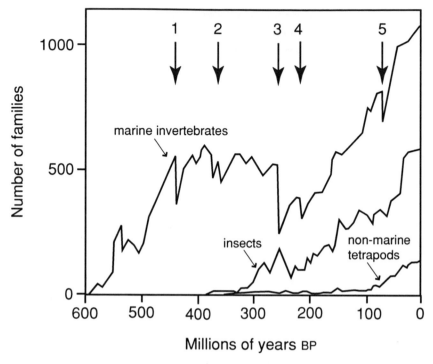

Figure 1.5. Diversity profiles for families of marine invertebrates, insects and non-marine tetrapods during the past 600 million years (latest Precambrian and Phanerozoic). Positions of the Big Five mass extinctions indicated by arrows (1, end-Ordovician; 2, Late Devonian; 3, end-Permian; 4, end-Triassic; 5, end-Cretaceous). After Benton (2001). BP, Before Present.

of Figure 1.5. Drops in diversity do, however, occur at the ends of the Permian, Triassic and Cretaceous periods.

The shapes of these diversity curves depend on the numbers of originations and extinctions from one interval of time to the next. The diversity for a given time interval (D^{x+1}) equals the diversity of the preceeding interval (D^x) plus the number of originations (O) minus the number of extinctions (E): $D^{x+1} = D^x + O - E$. If originations exceed extinctions then diversity increases; if extinctions exceed originations, diversity decreases. Conspicuous declines in diversity generally correlate with above average numbers of extinctions rather than below average numbers of originations, although low origination rate has occasionally been invoked to explain major extinctions. A recent analysis (Foote, 2000) has found that extinction has a greater role than

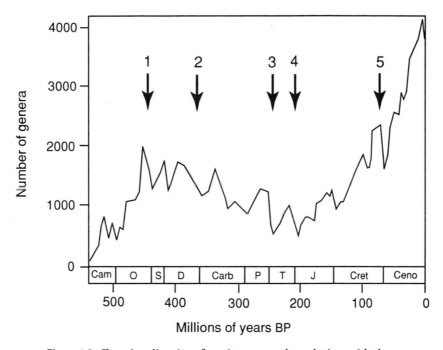

Figure 1.6. Changing diversity of marine genera through time with the positions of the Big Five mass extinctions indicated by arrows (1, end-Ordovician; 2, Late Devonian; 3, end-Permian; 4, end-Triassic; 5, end-Cretaceous). Geological period abbreviations: Cam, Cambrian; O, Ordovician; S, Silurian; D, Devonian; Carb, Carboniferous; P, Permian; T, Triassic; J, Jurassic; Cret, Cretaceous; Ceno, Cenozoic. After Newman (2001) using the data of J. J. Sepkoski.

origination in shaping the Palaeozoic diversity pattern, whereas the reverse is true for the post-Palaeozoic when origination had the more important role. The reason for this contrast has yet to be explained but may be connected with some aspect of ecosystem structure in the Palaeozoic that increased the impact of physical perturbations on life (Foote, 2000).

Extinction intensity in Phanerozoic marine animals can be quantified using the extinction metrics mentioned above. This has led to three important claims: (1) background extinction has declined through the Phanerozoic (Raup and Sepkoski, 1982); (2) five times of increased extinction (mass extinctions) stand out above the general background (Raup and Sepkoski, 1982); and (3) a periodicity of 26–30 Ma characterizes extinctions during the past 250 Ma (Raup and Sepkoski, 1984).

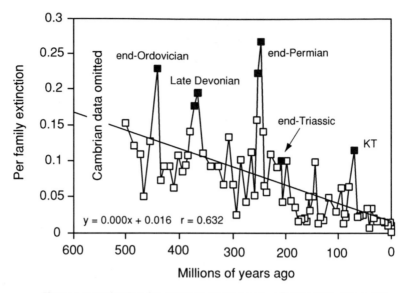

Figure 1.7. Extinction intensity, measured as per family extinctions, for marine and terrestrial organisms using the range information in Benton (1993), omitting angiosperms and pre-Ordovician data. The linear regression shows how extinction intensity has declined through time, and the infilled squares correspond to the Big Five mass extinctions previously recognized by Raup and Sepkoski (1982) for marine families only. Note that the end-Triassic mass extinction does not stand out especially from the general level of high extinction pertaining at this time (see Chapter 4).

Decline in background extinction

When calculated as extinction rate, average extinction intensity exhibits a progressive and steady decline from the Cambrian to the present day. For families, Raup and Sepkoski (1982) calculated a decline in family extinction rate from 4.6 per Ma in the lower Cambrian to about two per million years in the Holocene. Declining extinction is also evident in the family-level data of Benton (1993) which includes terrestrial as well as marine organisms (Figure 1.7). The explanation for decreasing background extinction, which is also apparent in marine genera, is a contentious issue. Does it indicate that taxa have in some way become more extinction resistant through time, perhaps signifying progressive adaptive improvements, or is the pattern an artefact of the nature of the data?

Flessa and Jablonksi (1985) drew attention to the possibility that declining extinction intensity may be a result of taxonomic structure.

The numbers of species per genus, and of species per family, are both known to have increased through geological time. This is probably a straightforward consequence of the branching structure of the tree of life – with time, fewer and fewer taxa above species-level originate and most new taxa comprise species that fit within already existing genera. The increasing number of species per higher taxon through the Phanerozoic means that for a given number of species becoming extinct, progressively fewer genera and families will suffer extinction. A simple model (Figure 1.8) illustrates how this artefact is produced. It has yet to be determined whether all of the decrease in background extinction is due to taxonomic artifacts.

Mass extinctions

Raup and Sepkoski (1982) in their analysis of extinction rates in marine families identified five times of particularly high extinction intensity. These events, which have become known as the Big Five mass extinctions, occurred at or near the ends of the Ordovician, Devonian, Permian, Triassic and Cretaceous periods. Their positions at the ends of geological periods are not coincidental – the boundaries between periods of geological time were originally recognized on the basis of conspicuous changes in fossil faunas and floras. Newer data on marine genera have typically supported the existence of the Big Five mass extinctions. However, two of the mass extinctions seem to be compound events, the Late Ordovician event containing two separate pulses of extinction and the Late Devonian containing five pulses. There is also evidence for a minor pulse preceding the main pulse of the Late Permian mass extinction. Furthermore, the question has not yet been settled whether mass extinctions are truly different from background extinctions. Perhaps there is a continuity in extinction magnitude, with mass extinctions simply representing high end members? A study published by Wang in 2003 has clarified the argument by pointing out the distinction between continuity of extinction intensity, continuity of cause and continuity of effect. While continuity of intensity seems evident from the available data, lack of continuity of cause (see Chapter 5) and of effect (see Chapter 6) may still separate mass from background extinctions.

Do non-marine animals and plants show the same five mass extinctions? Terrestrial vertebrates (tetrapods) had not evolved at the time of the Late Ordovician mass extinction and were rare when the mass extinction struck in the Late Devonian. However, they too show

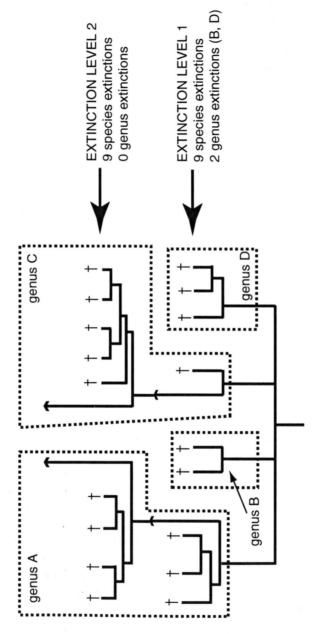

Figure 1.8. Model showing how declining extinction rates of taxa through time above the species-level may be the result of taxonomic structure causing geologically older genera to contain fewer species than younger genera. In this hypothetical example, the extinction (dagger) of an equal number of species, i.e. nine, results in a greater number of genus extinctions at the first than the second extinction level.

Table 1.1. *Animal groups suffering significant levels of extinction during the* 'Big Five' *mass extinctions*

Late Ordovician	Late Devonian	End Permian	Late Triassic	End Cretaceous
Trilobites	Trilobites	Trilobites*	Brachiopods	Foraminifera
Brachiopods	Brachiopods	Rostroconchs*	Ammonoids	Bryozoans
Nautiloids	Nautiloids	Echinoids	Gastropods	Corals
Crinoids	Ammonoids	Crinoids	Bivalves	Sponges
	Stromatoporoids	Blastoids*	Sponges	Echinoids
	Tabulate corals	Brachiopods	Marine reptiles	Brachiopods
	Rugose corals	Fenestrate bryozoans*	Terrestrial reptiles	Gastropods
	Crinoids	Ammonoids	Freshwater fishes	Bivalves
	Placoderm fishes	Rugose corals*	Insects	Ammonites*
		Terrestrial reptiles		Belemnites*
				Plesiosaurs*
				Mosasaurs*
				Pterosaurs
				Dinosaurs
				Marsupial mammals

*Indicates groups becoming totally extinct.
Based on Benton (1986).

high levels of extinction in the Late Permian, Late Triassic and Late Cretaceous. Insects exhibit a clear extinction at the end of the Permian but seem to have been less affected by the K–T event. While plants appear to have been less susceptible to mass extinctions (Chapter 3), the end Permian and to a lesser degree the K–T mass extinctions are detectable in the plant fossil record. Victims of major mass extinctions were clearly drawn from a wide spectrum of taxonomic groups (Table 1.1), animals and plants, marine and terrestrial. However, different taxonomic groups can show distinctly different patterns of diversity change at a mass extinction (Figure 1.9).

Estimates of the severity of these mass extinctions for species have been made using a technique called reverse rarefaction. The results suggest that 84–85 per cent of marine species became extinct in the

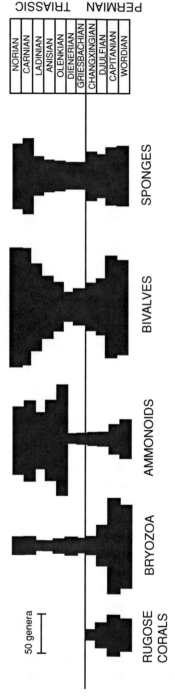

Figure 1.9. Contrasting patterns of clade diversity change for some selected marine groups in the Upper Permian and Triassic (after Erwin, 1994). The width of the black bars scales to generic diversity but their height has not been standardized despite variations in the durations of the stratigraphical stages (right). Rugose corals and bryozoans both declined towards the end of the Permian, the former becoming extinct but the latter surviving at a reduced diversity for the entire Triassic. Ammonoid and bivalve molluscs also show bottlenecks around the Permian–Triassic boundary but recovered more quickly than bryozoans. In the case of bivalves, generic diversity bottomed-out at a minimum value after the the main extinction level. Sponges seem to have been much less affected by the mass extinction.

Late Ordovician event, 79–83 per cent in the Late Devonian, 95 per cent in the Late Permian, 79–80 per cent in the Late Triassic, and 70–76 per cent in the Late Cretaceous (Jablonski, 1995).

Are there any differences between taxa becoming extinct during mass extinctions and those disappearing at other times? Boyajian (1991) was unable to find any differences in the longevities of families that were victims of mass versus background extinctions. There are, however, some marked differences in other biological traits. For example, Jablonski's study (1986) of Cretaceous and Palaeogene marine molluscs in the Gulf and Atlantic Coastal Plain of North America found that species with planktotrophic larvae (long-lived and capable of feeding) fared better than those with non-planktotrophic larvae (short-lived and reliant on parental provisioning) during times of background extinction. In contrast, there was no difference in survival between species possessing these two larval types during the K–T mass extinction.

While mass extinctions are important as sudden, intense 'cullings' of taxa that have disproportionate evolutionary consequences (Erwin, 2001), most species suffering extinction during the Phanerozoic did so at times of background rather than mass extinction. For the marine realm, simulations indicate that nearly 40 per cent of total species extinctions may have occurred during time intervals with per species extinction intensities of five per cent or less, compared to perhaps as little as 10 per cent at times of high intensity when species losses reached 50 per cent or more (Raup, 1991, 1995).

The long-term ecological effects of the Big Five mass extinctions varies. In terms of numbers of extinctions of marine families, the Late Ordovician and Late Devonian mass extinctions are very similar, but the former entailed only minor ecological changes whereas the latter brought about a radical restructuring of the many marine ecosystems (Droser et al., 2000). The taxonomic and ecological severity of mass extinctions may therefore be decoupled.

Extinction periodicity

Data on marine family extinctions for the Mesozoic and Cenozoic shows evidence of a periodic pattern of peaks in extinction intensity (Figure 1.10). This periodicity was first recognized by Raup and Sepkoski (1984) and analysed further by Sepkoski and Raup (1986). It has been the centre of considerable debate, both with regard to whether the pattern is real or an artefact, and what may have caused the periodicity.

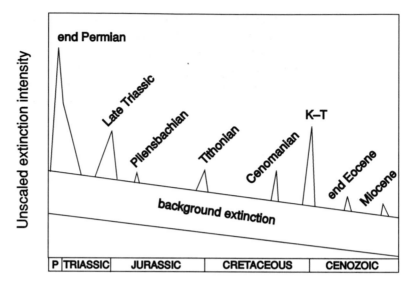

Figure 1.10. Apparent periodicity in extinction intensity for marine families during the Mesozoic and Cenozoic. Of the 10 extinction peaks predicted from a 26-million year periodicity, eight seem to match known extinction events recognized in the fossil record and labelled in the figure. After Sepkoski and Raup (1986).

Peaks of marine family extinction were calculated to occur at intervals of 26.2 Ma, commencing with the end-Permian mass extinction about 250 Ma ago (Sepkoski and Raup, 1986). This 26.2 million year periodicity should give 10 extinction peaks in the post-Palaeozoic. Eight of the 10 predicted peaks do indeed coincide with known extinction events in the fossil record, including the famous K–T event. However, at least one of these extinctions is of marginal veracity, the chronological match to others is not exact, and predicted extinctions in the Bajocian and Hauterivian stages are not evident to palaeontologists working in these time intervals. Agreement has yet to be reached about the 'reality' of extinction periodicity in the Mesozoic and Cenozoic, and as yet there is no indication of the same periodicity extending back into the Palaeozoic.

Extinction selectivity

Extinctions do not affect all taxa equally. Instead, both mass and background extinctions may exhibit selectivity according to various biological traits, taxonomic position and geographical factors (McKinney, 2001). Biological traits that tend to 'promote' extinction include large body size, dietary, thermal and other kinds of ecological specialism, low reproductive output and slow growth rate. Conversely, small generalist species with high fecundity and rapid growth rate are more likely to survive extinction. Rarity, both in terms of restricted geographical range and low population density, also makes taxa more prone to extinction. One reason that taxa with large body size are prone to extinction is that large size generally correlates with rarity – large organisms typically have small population sizes, meaning that the death of relatively few individuals can bring about extinction of the taxon. Among marine organisms, planktonic taxa sometimes suffer more than benthic taxa during times of mass extinction. This is true, for example, of the K–T extinction where planktonic foraminifera show higher levels of extinction than benthic foraminifera.

Some taxonomic groups routinely show higher extinction rates than others, reflected in the shorter mean durations of their constituent species. For example, ammonites have higher extinction rates than bivalve molluscs. With regard to geographical selectivity, tropical taxa commonly suffer disproportionately during mass extinctions compared to non-tropical taxa. A manifestation of intense tropical extinction is the decimation of reefs seen during mass extinctions (Chapter 4). Taxa inhabiting freshwater environments exhibit below average levels of extinction compared to marine species during the Late Devonian and K–T mass extinctions. Finally, extinction levels may vary across the globe; for example, marine molluscs in North America show higher levels of extinction at the end of the Cretaceous than do those from elsewhere in the world.

INTERPRETATION OF EXTINCTION PATTERNS AND PROCESSES

Evolutionary patterns are the products of evolutionary processes. A recurring issue afflicting public and creationist understanding of evolution is the failure to separate pattern from process. For example, descent with modification (an evolutionary pattern) is often inextricably linked with Natural Selection (an evolutionary process), although accepting the first does not demand acceptance of the second.

Evolutionary patterns also include the distribution of morphological characters within taxa, which is the basis for inferring phylogeny and reconstructing evolutionary trees, and the distribution of fossil taxa in time. While it is possible to read patterns directly from the fossil record, the patterns we see are not entirely due to evolution. 'Noise', including sampling artefacts, must be factored out before we can perceive the evolutionary 'signal'. In contrast, the processes creating the patterns we see in the fossil record are historical – they happened in the distant geological past. Extinction processes, including identification of the factors bringing about the extinction of a species, are consequently more difficult to study. That determining the cause of extinctions in fossil species is a hazardous business should be clear when we consider that the reasons for the extinctions of species in historical times are often uncertain. Likewise, the cause is as yet unknown for an abrupt population decline in London and elsewhere in western Europe of the House Sparrow (*Passer domesticus*) that threatens local extinction of this most easily observed and studied species (Hole *et al.*, 2002). Nevertheless, the geological record sometimes provides strong circumstantial evidence for the factors driving ancient extinctions, notably mass extinctions. Interpreting extinction processes in the history of life is certainly not beyond the scope of legitimate scientific study.

Interpreting extinction patterns

The fossil record samples a mere fraction of the life that has existed on Earth. Animal species without mineralized hard skeletons are absent or scarce in the fossil record, and some diverse present day phyla (e.g. nematode worms) have a pitiful fossil record. An estimated 250 000 fossil species are known, compared with a present day diversity of perhaps 10–50 million species (Pimm *et al.*, 1995). If it is true that some 99 per cent of Earth's biota is now extinct, as suggested by fossil species durations, then the total number of species to have inhabited the Earth through geological time may have exceeded 1000 million, and the proportion known from the fossil record could be as little as 0.02 per cent. We cannot know for sure whether this tiny fraction of fossilized species is representative of the biosphere as a whole with respect to extinction patterns, but this is a reasonable working hypothesis in the absence of evidence to the contrary.

Databases of taxa and their ranges from which extinction patterns are derived have long been acknowledged to contain various errors and potential biases (Smith, 2001) – the way in which taxonomic structure may have produced a pattern of declining background extinction through time has already been discussed (p. 17). An important question is whether these errors and biases are randomly distributed through time and merely obscure underlying patterns or are distributed non-randomly and create an artificial pattern of their own. Substantial range and taxonomic corrections may have little effect on broad evolutionary patterns (e.g. Sepkoski 1993; Adrain and Westrop 2000). However, the fossils available for reconstructing these patterns depend on the availability of fossil-bearing sedimentary rocks of appropriate age. Unfortunately, the stratigraphical record is not uniform – sedimentary rock outcrop area is greater for some intervals of geological time than it is for others, usually correlating with global sea-level because flooding of continental shelves during times of high sea-level causes sediments to accumulate over wide areas. For post-Palaeozoic shallow marine invertebrates, Smith (2001) found that per genus extinction peaked every 27 Ma, correlating with times when rock outcrop area was either maximal or minimal. These times correspond respectively to the tops of stacked transgressive system tracts and the bases of second-order stratigraphical sequences. These are positions at which first (and last) appearances can be predicted to cluster simply as a result of sequence architecture (Holland, 1995). A reasonable match between the 27-Ma periodicity found by Smith (2001) and Raup and Sepkoski's (1984) 26.2-Ma periodicity mentioned above is evident here.

It is not uncommon for palaeontologists to deny the importance of a mass extinction for a particular taxonomic group on the grounds that the group was in decline anyway before the mass extinction. When evaluating arguments of this kind one must bear in mind not only the Signor–Lipps Effect (see above) but also the even probability that the diversity of a particular clade will be decreasing at the time of a mass extinction purely by chance in the context of the waxing and waning of clades through time. Nonetheless, some groups previously considered to have perished during a mass extinction are now thought to have become extinct at an earlier time. Perhaps the best example are the rudistid bivalves (Figure 1.2). These aberrant reef builders ranged well into the terminal Maastrichtian stage of the Cretaceous but appear to have disappeared half a million years before the K–T mass extinction (although this claim has recently been challenged by Steuber et al., 2002).

Conclusions in experimental science are not accepted until the experiment has been repeated and the results replicated in more than one laboratory. An equivalent procedure in palaeontology would be to resample a geological section and see whether the same species are found with the same stratigraphical ranges when the new samples are analysed blindly (i.e. without knowledge of their relative stratigraphical levels) by independent groups of palaeontologists. This has been done for a K–T boundary section at El Kef in Tunisia. The aim of the El Kef blind test was to see which of two assertions made previously about the extinction pattern of planktonic foraminifera was the more correct: (1) mass extinction at the K–T boundary; or (2) stepwise extinction of species leading up to the K–T boundary. No agreement was reached between the four teams of palaeontologists involved – some favoured the first pattern and others the second. Lack of taxonomic consistency and the inability to recognize species reworked into higher stratigraphical levels were thought to explain the lack of consensus. A subsequent study (Arenillas *et al.*, 2000) outlined the problems associated with the El Kef blind test and, on the basis of new sampling, favoured the mass extinction hypothesis, with a loss of 74 per cent of species at the K–T boundary itself.

Interpreting extinction processes

As noted above, the causes of extinction for individual species in the fossil record may be difficult to pinpoint or verify. However, large-scale events that drive many species simultaneously to extinction can leave strong biological and geological signatures, allowing a testable hypothesis of the extinction process to be proposed. Such hypotheses are complicated by the fact that they may: (1) comprise a long chain of events, from an ultimate trigger (e.g. asteroid impact) to a final kill mechanism (e.g. starvation); and/or (2) involve the action of two or more independent factors, often with a short-lived perturbation ('straw that broke the camel's back', Zinsmeister, 1998) acting on a biota suffering longer term stress for other reasons (Chapter 4).

Proposed processes for species extinction are legion. They can be subdivided into biotic and abiotic, with the caveat that causal chains may involve both. As mentioned above, Darwin (1859) favoured competition between species as the main reason for extinction. Small-scale ecological experiments in closed systems, such as the classic experiments of Gause (1934) using species of the protist *Paramecium* competing for the same food, have demonstrated the existence of competitive exclusion and its potential for extinction of one species by another. It is debatable, however, to what extent this extinction mechanism operates in open, complex and dynamic natural environments where physical disturbance can prevent competitive exclusion (small islands may provide an exception). Biological invasions following the breakdown of biogeographical barriers, bringing species into contact for the first time, have usually not resulted in extinction of the resident species (Vermeij, 1991; Davis, 2003). Similarly, the diversification of a clade of competitively superior bryozoans (cheilostomes) did not cause an increase in extinction rates in another bryozoan clade (cyclostomes) that had been established for longer but generally lost competitive encounters for living space (McKinney and Taylor, 2001).

Predation and disease are two other biotic causes of extinction in the fossil record for which there is some evidence of potency in historic times (e.g. Maynard Smith, 1989). Ozanne and Harries (2002) recorded an increase in the proportion of inoceramid bivalves showing evidence from shell damage of predation and/or predation immediately before their extinction. This prompted them to suggest that these biotic agents may have played a role in decimating populations and making them vulnerable to extinction. Relatively few comparable cases have been argued. With rare exceptions, competition, predation and disease are in general difficult to detect in the fossil record. This does not mean that ancient extinctions were never caused by biotic factors; rather that we have few ways to detect them and to evaluate their importance in extinction.

Abiotic, non-biological factors driving species to extinction have been the subject of considerable interest and debate among palaeontologists. This is because abiotic mechanisms generally provide more readily testable causal hypotheses of extinction, particularly for mass or other major extinctions where a strong biological and/or geological signature may point to a causal process. Factors commonly implicated in extinction hypotheses include volcanism, bolide impacts, sea-level rise, sea-level fall, anoxia, global warming, global cooling and changing continental configuration. All of these may leave conspicuous traces in the geological record. For example, flood basalt provinces provide evidence of major volcanism; impacting enriches accumulating sediments in iridium and generates shocked quartz; sea-level rise and fall causes expansion and contraction respectively of the areal spread of shallow marine sediments on continental shelves; anoxia is indicated by deposition of carbonaceous black shales lacking fossils of animals living on or beneath the sediment surface; global warming allows reefs and related tropical environments to enlarge their latitudinal distribution; global cooling may leave glacial features such as ice-scoured rock surfaces; and changes in continental configurations are indicated by a wealth of geological evidence, including palaeomagnetic data.

Analysis of variations in the isotopic compositions of sediments and fossils through time are providing new insights into environmental change. Associated with most mass extinctions are negative excursions in $\delta^{13}C$, a measure of the proportion of ^{13}C to ^{14}C. The interpretation of such isotopic spikes varies – collapse in biological productivity is often invoked, an expected consequence of biotic decimation. Alternatively, spikes may indicate the release of large quantities of carbon dioxide, for example from gas hydrates, which have very light δ13C values, in the sediments of the continental shelf or massive volcanism, both potential causal processes in mass extinctions.

Linking any particular extinction event with one of these abiotic causal mechanisms is seldom straightforward. For one thing it demands good correlation – the extinction and the geological signature must coincide exactly. Keller and coworkers' (2003) detailed study of the Chicxulub crater thought by many geologists to have been formed by the impact of the asteroid that led to the K–T mass extinction has suggested that cratering occurred about 300 000 years before the mass extinction. Secondly, unlike a controlled laboratory experiment, the dynamic Earth has not varied one environmental factor while holding all of the others constant. Isolating the effects of a single hypothesized extinction mechanism can be very difficult. Furthermore, complex interactions can occur between factors. For example, global cooling may act directly by eliminating warm-adapted species. Locking of water in the icecaps formed by intense cooling will also cause sea level to fall, diminishing the habitable area available to shallow marine organisms and at the same time reducing the outcrop area of shallow water marine sediments preserved in the geological record. Cooling and smaller habitat area are both potential sources of extinction, while reduction in outcrop area from which geologists can sample fossils will further amplify the pattern of extinction seen in the fossil record. Such a combination of factors is thought to have driven the first pulse of extinction at the end of the Ordovician.

A good example of how biological selectivity (see above, p. 22) can provide clues to the processes causing mass extinctions is provided by Smith and Jeffery's (1998) study of sea urchin (echinoid) extinction at the K–T boundary. These authors undertook a global taxonomic study of echinoids and looked for correlations between different biological traits and extinction/survival. While many traits showed no relationship, feeding strategy was well correlated with extinction probability – omnivorous regular echinoids suffered less than specialist herbivores, and deposit feeding irregular echinoids equipped with pencillate tube feet allowing them to process very fine sediments for food fared better than other deposit feeding irregulars. This extinction selectivity supports the idea that a drop in nutrient supply was a major factor driving the K–T mass extinction. That this nutrient drop was not a short-term effect is indicated by the small size of nearly all surviving echinoids for a considerable interval of time after the extinction event.

Various attempts have been made by different scientists to advocate definitive roles for impacting, sea-level rises and falls, major volcanism or climatic changes in mass extinction. So far, however, none have been able to demonstrate sufficiently exact correlations in time between any of these factors and known extinction events to support a single causal mechanism for mass extinctions in the fossil record.

CONCLUSIONS

Just as the fate of all individuals is death, so that of all species is extinction. Fossils provide ample evidence for the former existence of a multitude of extinct species on Earth. Many of these extinct species belong to major taxonomic groups that have left no survivors, such as trilobites and ammonites. Reading and interpreting the record of extinction presents palaeontologists with a major challenge. Ever since Cuvier 200 years ago provided convincing evidence for the reality of ancient extinctions, palaeontologists have grappled with such issues as the role of extinction in shaping the tree of life, and whether extinctions are the result of sudden, exceptional events (catastrophism) or gradual, everyday processes (uniformitarianism). With the advent of large databases of fossil distributions through time, more sophisticated forms of analysis (e.g. Newman and Palmer, 2003) and the development of new geological methods of investigation (e.g. isotope geochemistry), patterns and processes of extinction are becoming much more amenable to investigation.

The fossil record shows us that the biosphere often takes an enormous length of time to recover from a major extinction event (Chapter 6; Erwin 2001). Peaks of taxonomic origination follow extinction peaks by roughly 10 million years for both mass and background extinction (Kirchner and Weil, 2000). Biotas immediately succeeding mass extinctions are not only lower in diversity but may be relatively rare and comprise weedy, generalist species often of small body size (Hansen *et al.*, 1987; Twitchett 2001). Recovery patterns can show significant geographical variation, and the rapidity of recovery may differ from one major extinction to the next. The post-extinction world can be populated by an unusual biota, including anachronistic taxa (Chapter 4) given a rare opportunity to flourish in the absence of taxa that normally dominate.

ACKNOWLEDGMENTS

I thank Bill Schopf for suggesting the symposium from which this chapter arose and for inviting me to spend a period in residence as a CSEOL Visiting Fellow at UCLA where he, Jane Shen-Miller and Richard Mantonya all helped in various ways to make my stay productive and enjoyable.

FURTHER READING

Alvarez, W. T.rex and the crater of doom. Princeton: Princeton University Press, 1997.

Archibald, J. D. Dinosaur Extinction and the End of an Era: What the Fossils Say. New York: Columbia University Press, 1996.

Benton, M. J. Diversity and extinction in the history of life. Science 1995, 268: 52–58.

Benton, M. J. When Life Nearly Died. The Greatest Mass extinction of All Time. London: Thames & Hudson, 2003.

Briggs, D. E. G. and Crowther, P. R. (Eds.), Palaeobiology II. Oxford: Blackwell, 2001.

Courtillot, V. Evolutionary Catastrophes: The Science of Mass Extinction. Cambridge: Cambridge University Press, 1999.

Donovan, S. K. (Ed.). Mass Extinctions: Processes and Evidence. London: Belhaven, 1989.

Ehrlich, P. and Ehrlich, A. Extinction: The Causes and Consequences of the Disappearance of Species. New York: Random House, 1981.

Elliott, D. K. (Ed.) Dynamics of Extinction. New York: Wiley, 1986.

Erwin, D. H. The Great Paleozoic Crisis: Life and Death in the Permian. New York: Columbia University Press, 1993.

Hallam, A. and Wignall, P. B. Mass Extinctions and their Aftermath. Oxford: Oxford University Press, 1997.

Mass extinctions and sea-level changes. Earth-Science Reviews 48 (1999), 217–50.

Kauffman, E. G. and Walliser, O. H. (Eds.). Extinction Events in Earth History. New York: Springer, 1990.

Larwood, G. P. (Ed.). Extinction and Survival in the Fossil Record. Oxford: Clarendon Press, 1988.

Lawton, J. H. and May, R. M. (Eds.). Extinction Rates. Oxford: Oxford University Press, 1995.

MacLeod, N. and Keller, G. *Cretaceous–Tertiary Mass Extinctions: Biotic and Environmental Changes*. Norton: New York, 1996.

Martin, P. .S. and Klein, R. G. (Eds.). *Quaternary Extinctions: A Prehistoric Revolution*. Tucson: University of Arizona Press, 1984.

McGhee, G. R. Jr. *The Late Devonian Mass Extinction*. New York: Columbia University Press, 1996.

Newman, M. E. J. and Palmer, R. G. *Modeling Extinction*. Oxford: Oxford University Press, 2003.

Nitecki, M. H. (Ed.). *Extinctions*. Chicago: University of Chicago Press, 1984.

Raup, D. M. *The Nemesis Affair: A Story of the Death of Dinosaurs and the Ways of Science*. London: Norton, 1986.

Raup, D. M. *Extinction: Bad Genes or Bad Luck*. New York: Norton, 1991.

Stanley, S. M. *Extinction*. New York: Scientific American Books, 1987.

Wignall, P. B. Large igneous provinces and mass extinctions. *Earth-Science Reviews* **53** (2001), 1–33.

REFERENCES

Alvarez, L. W., Alvarez, W., Asaro, F. and Michel, H. V., 1980. Extraterrestrial cause for the Cretaceous–Tertiary extinction: experimental results and theoretical interpretation. *Science* **208**: 1095–1108.

Adrain, J. M. and Westrop, S. R., 2000. An empirical assessment of taxic paleobiology. *Science* **289**: 110–112.

Arenillas, I., Arz, J. A., Molina, E. and Dupuis, C., 2000. An independent test of planktic foraminiferal turnover across the Cretaceous/Paleogene (K/P) boundary at El Kef, Tunisia: catastrophic mass extinction and possible survivorship. *Micropaleontology* **46**: 31–49.

Benton, M. J., 1986. The evolutionary significance of mass extinctions. *Trends in Ecology and Evolution* **1**: 127–130.

  1993. *The Fossil Record 2*. London: Chapman & Hall.

  2001. Biodiversity through time. In: D. E. G. Briggs and P. R. Crowther (Eds.), *Palaeobiology II*. Oxford: Blackwell, pp. 212–220.

Boyajian, G. E., 1991. Taxon age and selectivity of extinction. *Paleobiology* **17**: 49–57.

Cuvier, G., 1812. *Recherches sur les ossemens fossiles de quadrupèdes, ou l'on rétablit les caractères de plusieurs espèces d'animaux que les révolutions du globe paroissent avoir détruites*. Paris: Deterville, 4 vols.

Darwin, C., 1859. *The Origin of Species by Means of Natural Selection or the Preservation of Favoured Races in the Struggle for Life*. London: Murray.

Davis, M. A., 2003. Biotic globalization: does competition from introduced species threaten biodiversity? *BioScience* **53**: 481–489.

Droser, M. L., Bottjer, D. J., Sheehan, P. M. and McGhee, G. R., Jr, 2000. Decoupling of taxonomic and ecologic severity of Phanerozoic marine mass extinctions. *Geology* **28**: 675–678.

Erwin, D. H., 1994. The Permo-Triassic extinction. *Nature* **367**: 231–236.

　2001. Lessons from the past: biotic recoveries from mass extinctions. *Proceedings of the National Academy of Sciences* **98**: 5399–5403.

Fara, E., 2001. What are Lazarus taxa? *Geological Journal* **36**: 291–303.

Flessa, K. W. and Jablonksi, D., 1985. Declining Phanerozoic background extinction rates: effect of taxonomic structure? *Nature* **313**: 216–218.

Foote, M., 2000. Origination and extinction components of taxonomic diversity: Paleozoic and post-Paleozoic dynamics. *Paleobiology* **26**: 578–605.

Gause, G. J., 1934. *The Struggle for Existence*. Baltimore: Williams and Wilkins.

Hansen, T., Farrand, R. B., Montgomery, H. A., Billman, H. G. and Blechschmidt, G., 1987. Sedimentology and extinction patterns across the Cretaceous–Tertiary boundary interval in East Texas. *Cretaceous Research* **8**: 229–252.

Hole, D. G., Whittingham, M. J., Bradbury, R. B. *et al.*, 2002. Widespread local house-sparrow extinctions. *Nature* **418**: 931.

Holland, S. M., 1995. The stratigraphic distribution of fossils. *Paleobiology* **21**: 92–109.

Jablonski, D., 1986. Background and mass extinctions: the alternation of macroevolutionary regimes. *Science* **231**: 129–133.

　1995. Extinctions in the fossil record. In: J. H. Lawton and R. M. May (Eds.), *Extinction Rates*. Oxford: Oxford University Press, pp. 25–44.

Jackson, J. B. C., 2001. What was natural in the coastal oceans? *Proceedings of the National Academy of Sciences* **98**: 5411–5418.

Larwood, G. P. (Ed.) 1988. *Extinction and Survival in the Fossil Record*. Oxford: Clarendon Press.

Lyell, C. (1830). *Principles of Geology*, vol. 1. London: John Murray.

Keller, G., Adatte, T. and Stinnesbeck, W., 2003. The non-smoking gun. *Geoscientist* **13**: 8–11.

Kirchner, J. W. and Weil, A., 2000. Delayed biological recovery from extinctions throughout the fossil record. *Nature* **404**: 177–180.

Maynard Smith, J., 1989. The causes of extinction. *Philosophical Transactions of the Royal Society of London* **B325**: 241–252.

McKinney, F. K. and Taylor, P. D., 2001. Bryozoan generic extinctions and originations during the last one hundred million years. *Palaeontologia Electronica* **4** (1): Article 3, 26 pp. http://palaeo-electronica.org/2001_1/bryozoan/issue1_01.htm

McKinney, M. L., 2001. Selectivity during extinctions. In: D. E. G. Briggs and P. R. Crowther (Eds.), *Palaeobiology II*. Oxford: Blackwell, pp. 198–202.

Newman, M., 2001. A new picture of life's history on Earth. *Proceedings of the National Academy of Sciences* **98**: 5955–5956.

Newman, M. E. J. and Palmer, R. G. 2003. *Modeling Extinction*. New York: Oxford University Press.

Nicholson, H. A., 1879. *A Manual of Palaeontology*, vols. 1 and 2. Edinburgh: Blackwood.

Ozanne, C. R. and Harries, P. J., 2002. Role of predation and parasitism in the extinction of inoceramid bivalves: an evaluation. *Lethaia* **35**: 1–19.

Phillips, J., 1860. *Life on the Earth: its Origins and Succession.* Cambridge: Macmillan.

Pimm, S. L., Russell, G. J., Gittleman, J. L. and Brooks, T. M., 1995. The future of biodiversity. *Science* **269**: 347–350.

Rampino, M. R. and Adler, A. C., 1998. Evidence for abrupt latest Permian mass extinction of foraminifera: results of tests for the Signor–Lipps effect. *Geology* **26**: 415–418.

Raup, D. M. 1986. *The Nemesis Affair: A Story of the Death of Dinosaurs and the Ways of Science.* New York: Norton.

Raup, D. M., 1991. A kill curve for Phanerozoic marine species. *Paleobiology* **17**: 37–48.

1995. The role of extinction in evolution. In: W. M. Fitch and F. J. Ayala (Eds.), *Tempo and Mode in Evolution.* Washington: National Academy Press, pp. 109–124.

Raup, D. M., and Sepkoski, J. J., Jr, 1982. Mass extinctions in the marine fossil record. *Science* **215**: 1501–1503.

1984. Periodicity of extinctions in the geologic past. *Proceeding of the National Academy of Sciences* **81**: 801–805.

Rudwick, M. J. S., 1997. *Georges Cuvier, Fossil Bones, and Geological Catastrophes.* Chicago: University of Chicago Press.

Schindel, D. E., 1982. Resolution analysis: a new approach to the gaps in the fossil record. *Paleobiology* **8**: 340–353.

Sepkoski, J. J., Jr, 1981. A factor analytic description of the Phanerozoic marine fossil record. *Paleobiology* **7**: 36–53.

Sepkoski, J. J., Jr, 1993. Ten years in the library: new data confirm paleontological patterns. *Paleobiology* **19**: 43–51.

Sepkoski, J. J., Jr and Raup, D. M., 1986. Periodicity in marine extinction events. In: D. K. Elliott (Ed.), *Dynamics of Extinction.* New York: Wiley, pp. 3–36.

Signor, P. W., III and Lipps, J. H., 1982. Sampling bias, gradual extinction patterns and catastrophes in the fossil record. *Geological Society of America Special Paper* **190**: 291–296.

Smith, A. B., 2001. Large-scale heterogeneity of the fossil record: implications for Phanerozoic biodiversity studies. *Philosophical Transactions of the Royal Society, Series B* **356**: 351–367.

Smith, A. B. and Jeffery, C. H., 1998. Selectivity of extinction among sea urchins at the end of the Cretaceous period. *Nature* **392**: 69–71.

Steuber, T., Mitchell, S. F., Buhl, D., Gunter, G. and Kasper, H. U., 2002. Catastrophic extinction of Caribbean rudist bivalves at the Cretaceous–Tertiary boundary. *Geology* **30**: 999–1002.

Twitchett, R. J., 2001. Incompleteness of the Permian–Triassic fossil record: a consequence of productivity decline? *Geological Journal* **36**: 341–353.

Vermeij, G. J., 1991. When biotas meet: understanding biotic interchange. *Science* **253**: 1099–1104.

Wang, S. C., 2003. On the continuity of background and mass extinction. *Paleobiology* **29**: 455–467.

Wignall, P. B. and Benton, M. J., 1999. Lazarus taxa and fossil abundance at times of biotic crisis. *Journal of the Geological Society, London* **156**: 453–456.

Zinsmeister, W. J., 1998. Discovery of fish mortality horizon at the K–T boundary of Seymour Island: re-evaluation of events at the end of the Cretaceous. *Journal of Paleontology* **72**: 556–571.

J. WILLIAM SCHOPF

Department of Earth and Space Sciences, Institute of Geophysics and Planetary Physics (Center for the Study of Evolution and the Origin of Life), and Molecular Biology Institute, University of California, Los Angeles, USA

2

Extinctions in life's earliest history

GEOLOGICAL TIME

All of geological time – the total history of the Earth – is divided into two great Eons, the Precambrian and the Phanerozoic (Figure 2.1). The Precambrian Eon is the older and much the longer of the two, extending from when the planet formed, about 4500 million years (Ma: mega anna) ago, to the appearance of fossils of hard-shelled invertebrate animals such as trilobites and various kinds of shell-bearing molluscs at about 550 Ma ago. The eon is composed of two Eras, the older Archean Era (from the Greek *archaios*, ancient) that spans the time from 4500 to 2500 Ma ago; and the younger Proterozoic Era (the era of earlier life, from the Greek *proteros*, earlier, and *zoe*, life) that extends from 2500 Ma ago to the close of the Precambrian.

The younger and shorter eon is the Phanerozoic (the eon of evident life, from the Greek *phaneros*, evident or visible, and *zoe*). The Phanerozoic Eon encompasses the most recent roughly 550 Ma of Earth history and is divided into three eras (from oldest to youngest, the Palaeozoic, Mesozoic, and Cenozoic Eras; Figure 2.1), each subdivided into shorter segments known as geologic periods. The oldest such period of the Palaeozoic Era (and, consequently, of the Phanerozoic Eon), spanning the time from about 550 to roughly 500 Ma ago and named after Cambria, the Roman name for Wales where rocks of this age were first formally described, is known as the Cambrian Period. This explains why the earlier eon is named the Precambrian (formerly written as pre-Cambrian), an enormous thickness of rocks underlying, and thus older than all of those of the Phanerozoic, that until just a few decades ago

Extinctions in the History of Life, ed. Paul D. Taylor.
Published by Cambridge University Press. © Cambridge University Press 2004.

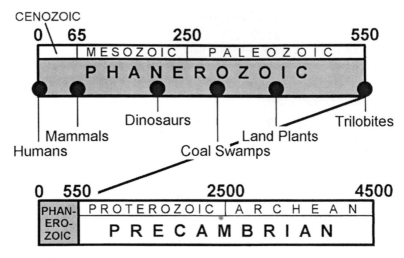

Figure 2.1. The major divisions of geological time. Ma, million years.

were universally regarded as lacking fossils. As best as anyone could tell, life of the Precambrian had left no trace.

The 'Standard View' of life's history

When we think of evolution, we think of the Phanerozoic history of life – the familiar progression from the first twig-like land plants to seed-producing flowering plants, from animals without backbones to fish, land-dwelling vertebrates, then birds and mammals. Yet Phanerozoic rocks are like the tip of an enormous iceberg, for they record only a brief late chapter, the most recent one-eighth, of a very much longer evolutionary story.

To visualize this, imagine that all 4500 Ma of geological time were condensed into a single 24-hour day (Figure 2.2). Evidence from the Moon and Mars tells us that for the first few hours of this day of Earth's existence its surface would have been uninhabitable, blasted by an intermittent stream of huge ocean-vaporizing meteorites. About 4:00 a.m., life finally gained a foothold. The oldest fossils, simple microbes, were entombed at about 5:30 a.m. Then, over an enormous segment of time, hardly anything happened. Although some of these early microbes (cyanobacteria) invented the process of oxygen-generating photosynthesis – a breakthrough that ultimately changed the world into one in which advanced organisms, such as ourselves, could survive and thrive – for millions upon millions of years, this early world would

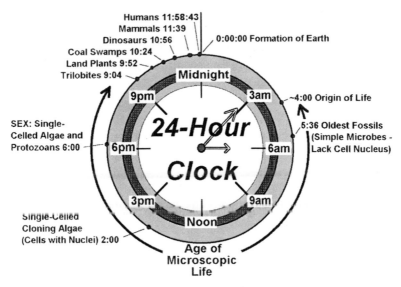

Figure 2.2. Geological time as a 24-hour day.

have seemed to us a bland, uninteresting place, inhabited only by micro-bial pond scum. Eventually, about 2:00 in the afternoon, the picture began to change: simple floating single-celled algae with cell nuclei and chromosomes appeared, but like their microbial ancestors, they also would have been too small to be seen without a microscope. About 6:00 p.m., as the sun began to set, a second major advance occurred, the appearance of sexual organisms – single-celled algae and protozoans still too small to be seen without a microscope, but microorganisms that possessed one of evolution's great inventions, sex, an innovation that would then spur all later evolutionary change. At about 8:30 in the evening, larger many-celled seaweeds entered the scene, and a few minutes later so did early-evolving jellyfish, worms, and hard-shelled trilobites.

The Precambrian, the period from the formation of the planet to the rise of shelled animals, spans 21 hours of this 24-hour geolog-ical day. The remaining three hours are left for the familiar Phanero-zoic evolutionary progression, the history of life recounted in texts and classrooms throughout the world. We humans arose only a few tens of seconds before midnight.

Life of the Phanerozoic, a world populated by large sexually repro-ducing plants and animals that thrive because their specialized organs (flowers, leaves, teeth, limbs) mesh so well with their surroundings,

has been studied actively since the early 1800s. And while such studies have shown that specialization is the key to the evolutionary success of Phanerozoic life, they have also revealed its downside, for if a
species is specialized for a given setting, it can be annihilated if the
setting changes. Thus, as is highlighted in the other chapters of this
book, the history of Phanerozoic life is punctuated by extinctions –
mostly local and hitting only a few species, but sometimes global and
devastating – and the rules of Phanerozoic evolution have come to be
well known: speciation (the formation of new species), specialization,
and extinction.

In place of the plants and animals of the Phanerozoic, the world
of the earlier Precambrian Eon was populated entirely, until near
the close of the eon, by small, simple, nonsexual microorganisms.
Rather than evolve at the pace normal for the large organisms of
the Phanerozoic, many of these microbes remained unchanged over
astonishingly long spans of time. Also, instead of being specialized for
local settings, members of the most successful of the ancient lineages –
cyanobacteria – are generalists that flourish in a remarkably wide range
of surroundings. So, rather than following the Phanerozoic rules of
evolution, life of the Precambrian tracked a decidedly different path:
speciation, generalization, and exceptionally long-term survival.

It is easy to understand why the Phanerozoic fossil record has
attracted such attention. Unlike Precambrian microbes, life of the
Phanerozoic has now been studied for more than two full centuries
and Phanerozoic fossils are large, striking, even awe-inspiring. That this
most recent 15 per cent of Earth history is the 'Age of Evident Life' is
more than just a handy moniker. But studies of the Precambrian – the
'Age of Microscopic Life' – have only just begun, and to some it is still a
new idea that during these early stages the game of life was played by
rules different from those of later evolutionary times. Yet as this chapter will show, the rules actually did change – evolution, itself, evolved.

Life in the Precambrian

How common fossils are of any biological group can be measured
by what are called taxonomic occurrences – the number of species
belonging to a given biological group that are known to be present
in officially recognized geological units. For instance, 10 species of
fossil cyanobacteria in each of three named geologic formations adds
up to 30 taxonomic occurrences. More than 6000 such occurrences of

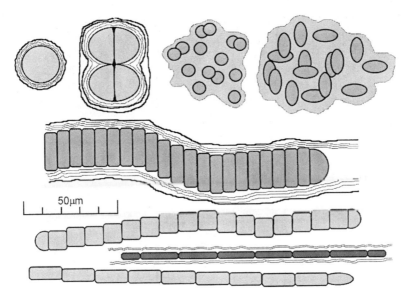

Figure 2.3. Cyanobacteria are of various shapes and sizes and are often surrounded by secreted layers of sticky mucilage.

fossil microorganisms, spread among nearly 800 geological deposits, are known from Precambrian rocks. Most of these are species of microbes, various types of bacteria and cyanobacteria, although microscopic single-celled algae are fairly common in deposits younger than about 1000 Ma in age. But the early fossil record is notably uneven. It extends to nearly 3500 Ma ago, to microbial fossils known from South Africa and northwestern Australia, but is meagre in rock units older than about 2200 Ma – mainly because rather few rock units from this time have survived to the present, and those that have survived have almost all been subjected to the fossil-destroying heat and pressure of mountain-building and other geological processes.

Cyanobacterial 'living fossils'

Among the thousands of Precambrian microscopic fossils known, by far the most common are various types of microbes, chiefly cyanobacteria (Figure 2.3), that are typically represented either by simple ball-shaped cells or by long stringy filaments that are either tubular and hollow or are made up of single-file rows of uniform cylinder-shaped cells. Abundant in finely layered nearshore mound-shaped deposits known

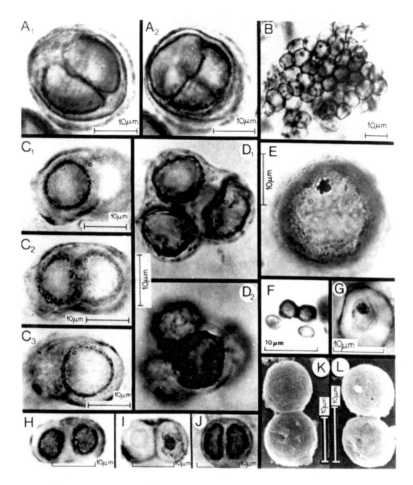

Figure 2.4. Optical photos (A through J) and scanning electron microscope pictures (K and L) of ball-shaped (chroococcacean) fossil cyanobacteria from the 850 million-years-old Bitter Springs Formation of central Australia.

as stromatolites, these ball- and string-like fossils are common also in siltstones and shales laid down in coastal lagoons and drying mud flats.

The ball-shaped fossils occur singly, in pairs, or in colonies of a few, hundreds, or even thousands of cells and are often surrounded by layers of wispy organic film, remnants of encasing mucilage envelopes. Their many kinds of colonies are well known among living bacteria – irregular masses, globe-shaped clusters, spheroidal rosettes, rectangular sheets, tiered cubes. The smallest-celled varieties are probably remnants of non-cyanobacterial microbes, but most of these fossils seem certain to be cyanobacteria, a widespread and remarkably successful early-evolved group of bacterial microorganisms (Figure 2.4). Indeed, many such fossils can be placed with confidence in a particular cyanobacterial taxonomic family (the Chroococcaceae), living members of which have the same cell sizes and shapes, and inhabit the same types of environments as their ancient, Precambrian relatives. One such example is shown in Figure 2.5F, the 1550-Ma-old four-celled colonial *Gloeodiniopsis*, a fossil that in all respects resembles the modern chroococcacean *Gloeocapsa* (Figure 2.5E).

The string-like filaments, whether tubular and hollow or segmented into distinct cells, are: unbranched; straight, curved, twisted, or coiled; and are often interlaced in felt-like mats that make up the stacked layers of stromatolites or the thin microbial veneers of mudflats. Narrow filaments are often long and sinuous, whereas thicker strands tend to be shorter, preserved as broken stubby fragments. The segmented strings are made up of cells uniform in size and shape except at their ends where they are capped by cells that commonly have a different shape. Some such fossils, particularly those that are quite narrow, are remnants of non-cyanobacterial microbes. But, like the ball-shaped forms that often occur in the same deposits, most are cyanobacteria, in this instance belonging to the modern cyanobacterial family Oscillatoriaceae (Figure 2.6). As like living members of this family, many of the fossil filaments were composed of two parts – a cylindrical string-like chain of cells (the 'trichome') encased by a loose- to tight-fitting hollow tube (the encompassing non-cellular 'sheath'). A good example is shown in Figure 2.5B, the 950-Ma-old sheath-enclosed trichome *Palaeolyngbya*, compared with *Lyngbya* (Figure 2.5A), its living cyanobacterial (oscillatoriacean) look-alike. In numerous deposits, only spaghetti-like masses of the tubular sheaths are preserved (Figure 2.7), partly because they become vacated and are left behind if their enclosed cellular trichomes move to new suroundings, but also because they are more resistant to decay than the cells of trichomic threads.

In addition to chroococcacean balls and oscillatoriacean strings, three other cyanobacterial families are fairly common in the early fossil record – the Pleurocapsaceae, Hyellaceae, and Entophysalidaceae. Pleurocapsaceans are egg-shaped colonial cyanobacteria that in some species form long close-packed stalks that radiate upward from the sea floor in pincushion-like groupings, and have changed little over time. Fossils known as *Eopleurocapsa* and *Paleopleurocapsa* are indistinguishable from the modern genus *Pleurocapsa*. Pincushion-like clumps of the stalked egg-shaped cells of *Polybessurus*, known from 770-Ma-old

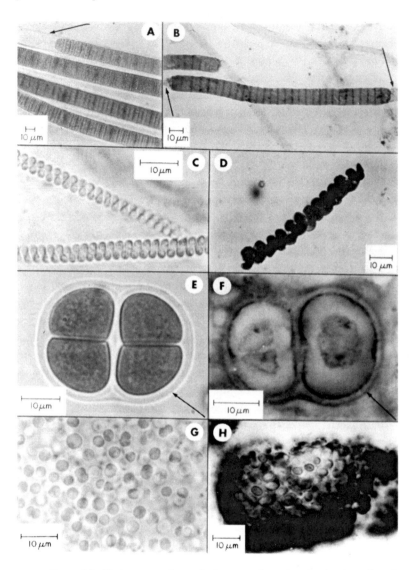

Figure 2.5. Modern cyanobacteria from northern Mexico (A, C, E, G) and Precambrian look-alikes (B, from the 950-million-years (Ma)-old Lakhanda Formation, and D, the 850 Ma-old Miroedikha Formation, both of Siberia; F, the 1550-Ma-old Satka Formation of Bashkiria; and H, the 2100-Ma-old Belcher Supergroup of Canada). (A) *Lyngbya*, compared with (B) *Palaeolyngbya*. (C) *Spirulina*, compared with (D) *Heliconema*. (E) *Gloeocapsa*, compared with (F) *Gloeodiniopsis*. (G) *Entophysalis*, compared with (H) *Eoentophysalis*.

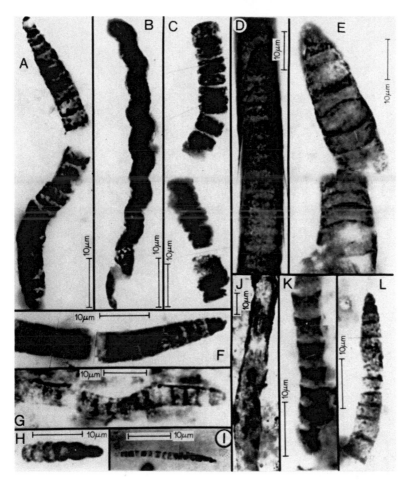

Figure 2.6. String-like (oscillatoriacean) cellular cyanobacterial fossils from the 850-million-year-old Bitter Springs Formation of central Australia.

stromatolites of South Australia and East Greenland, have the same morphology, reproduction, and pattern of growth as the pleurocapsacean *Cyanostylon* living today in coastal waters of the Great Bahama Bank, the same kind of environment inhabited by its fossil look-alike.

Hyellaceans are endoliths, cyanobacteria that etch tiny cavities into limestone pebbles, boulders, and stony pavements which they then inhabit, living within the outermost rock rind where sunlight penetrates. Like other cyanobacteria, fossil hyellaceans seem identical to living members of the group. For example, the Precambrian genus *Eohyella* was described by its discoverers as a 'compelling example of the

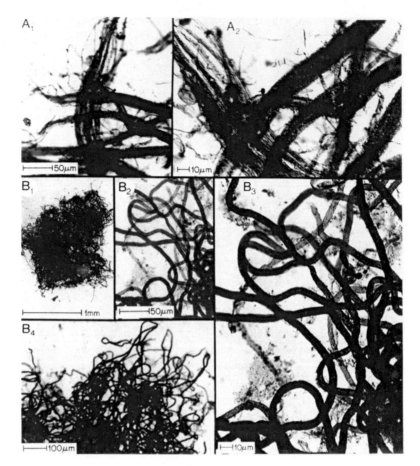

Figure 2.7. Flattened, originally tubular, cylindrical sheaths of filamentous (oscillatoriacean) cyanobacteria from the 850-million-year-old Miroedikha Formation of Siberia.

close resemblance between Proterozoic cyanobacteria and their modern counterparts' (Green *et al.*, 1988, p. 857), a fossil 'indistinguishable' from living *Hyella* of the eastern Caribbean (Green *et al.*, 1988, pp. 837–838).

Fossil–modern look-alikes are well known in yet another early-evolved family, the Entophysalidaceae, colonial cyanobacteria that have jelly-bean-shaped cells and form lumpy slime-embedded colonies on rocky substrates. An especially striking example is the Canadian 2100 Ma-old genus *Eoentophysalis* (Figure 2.5H) which is identical to modern *Entophysalis* (Figure 2.5G) in virtually all respects – in cell shape, colony form, the way its cells divide and grow, the finely layered stromatolitic structures it builds, the environments it inhabits, the make up of its microbial communities, even the way its cells break down when they die.

All five cyanobacterial families – the Chroococcaceae (balls), Oscillatoriaceace (cellular and tubular strings), Pleurocapsaceae (stalked eggs), Hyellaceae (cavity-inhabiting endoliths), and Entophysalidaceae (jelly beans) – include numerous examples of fossil-modern look-alikes; all have changed so little over time that they display what can be called 'status quo evolution'. Yet as impressive as these examples are, they reveal only a small part of the story. In fact, so many fossil–modern look-alikes have turned up in Precambrian deposits that when fossils new to science are found, it has become standard practice for their discoverers simply to name them after their present-day relatives. For example, fossilized microbes that match living cyanobacteria of the genus *Oscillatoria* have been named *Oscillatorites* ('related to *Oscillatoria*'), *Oscillatoriopsis* ('*Oscillatoria*-like'), and *Archaeoscillatoriopsis* ('ancient *Oscillatoria*-like'). Many workers have merely added the prefixes palaeo- ('old', as in *Palaeopleurocapsa* and *Palaeolyngbya*) or eo- ('dawn', as in *Eoentophysalis* and *Eohyella*) to names of living genera. Nearly 50 namesakes have been proposed by workers worldwide for fossil relatives of living cyanobacteria belonging to eight different taxonomic families (Figure 2.8).

The evidence is clear – diverse types of cyanobacteria changed little or not at all in shape, size, inhabited environments or general metabolic traits since they entered the scene literally billions of years ago. But as undeniable as this conclusion may be, it is nevertheless surprising, even startling. After all, as the history of Phanerozoic life well shows, 'living fossils' such as these are incredibly rare, curiosities that stick out and impress us because they are truly extraordinary. Yet during the Precambrian, the most successful of life's early branches is packed full of living fossils, of microbes that evolved at an almost imperceptibly sluggish rate. Why did cyanobacteria change so little over their exceedingly long history?

Status quo evolution

That cyanobacteria exhibit status quo evolution – stasis, a lack of evolutionary change – is a rather recent discovery, one that prior to the past few decades when the Precambrian fossil record first began to be uncovered had been entirely unforeseen. Everyone had expected that the standard rules of the Phanerozoic would apply equally to the Precambrian.

Cyanobacterial Fossil Namesakes

CHROOCOCCACEAE

Anacystis	Palaeoanacystis
Microcystis	Palaeomicrocystis
Gloeocapsa	Eogloeocapsa
Synechococcus	Eosynechococcus
Aphanocapsa	Eoaphanocapsa
Eucapsis	Eucapsamorpha

PLEUROCAPSACEAE

Pleurocapsa	Eopleurocapsa
Pleurocapsa	Palaeopleurocapsa

RIVULARIACEAE

Calothrix	Palaeocalothrix
Rivularia	Primorivularia

ENTOPHYSALIDACEAE

Entophysalis	Eoentophysalis

OSCILLATORIACEAE

Lyngbya	Palaeolyngbya
Spirulina	Palaeospirulina
Microcoleus	Eomicrocoleus
Phormidium	Eophormidium
Oscillatoria	Oscillatoriopsis
Schizothrix	Schizothropsis

NOSTOCACEAE

Nostoc	Palaeonostoc
Anabaena	Anabaenidium

SCYTONEMATACEAE

Plectonema	Eoplectonema
Scytonema	Palaeoscytonema

HYELLACEAE

Hyella	Eohyella

Namesakes coined by scientists in Brazil, Canada, China, India, Israel, Russia, USA

Figure 2.8. Fossil cyanobacteria (right columns) named after genera living today (left columns).

Early organisms would be smaller, simpler, perhaps less varied, but they were universally thought to have evolved in the same way and at the same rate as later life. Why is status quo evolution the rule, not the exception, for this early evolved, presumably primitive group of microbes that so dominanted Earth's earliest biosphere?

Of the various causes that can be suggested, three stand out. First, cyanobacteria do not reproduce sexually. As we will see below, sex greatly speeds evolution by serving up a huge supply of new combinations of genes, so its lack in cyanobacteria would have slowed this process. Second, cyanobacteria have huge populations (like those of most microorganisms, made up of billions, even trillions of individuals) and include many species that are essentially cosmopolitan (having been globally distributed by oceanic currents, winds, hurricanes, and the like), factors that would have delayed the spread of mutation-derived evolutionary change. Third – and most importantly – cyanobacteria can live almost anywhere; this versatility, unparalleled among other forms of life, is the key to their success.

CYANOBACTERIAL VERSATILITY

The versatility of cyanobacteria, both of individual species and of the group as a whole, is truly remarkable. As summarized in Figure 2.9, they live, even flourish, in almost total darkness to extreme brightness; in pure, salty, or the most saline waters; in acid hot springs or in lakes so alkaline that almost nothing else survives; in scalding pools or, in a dry state, above the boiling point of water; in icefields, frigid saline lakes, or hundreds of degrees below zero submersed in liquified helium or hydrogen; in the near-absence, presence, or huge over-abundance of oxygen or of carbon dioxide; in the driest place on earth, the Chilean Atacama Desert where it is claimed that rainfall has never been recorded; and even in the deadly ionizing radiation of a thermonuclear blast! Many can fix nitrogen gas, absorbing N_2 out of the atmosphere and building it into protein-forming amino acids. Provided with light, CO_2, a source of hydrogen (H_2O or, for some, even H_2S or H_2), and a few trace elements, cyanobacteria are champion biological pioneers, often the first to colonize newly formed volcanic islands. Versatile, resilient, and remarkably successful, the 'jack-of-all-trades' survival strategy of these early-evolved microbes – so different from that of the specialized plants and animals of the Phanerozoic – begs an explanation.

Light Intensity:	Extremely Dim (1 to 5 mEs^{-1}m^{-2}) Cultures		Normal Light (50 to 60 mEs^{-1}m^{-2}) Optimum Growth	Exceedingly Bright (>2,000 mEs^{-1}m^{-2}) Intertidal Zone
Salinity:	<0.001 to 0.1% Freshwater	3.5% Marine	27.5% Great Salt Lake	100 to 200% Salterns
Acidity/Alkalinity:	Acidic (pH 4) Hot Springs		Neutral (pH 7 to 9) Optimum Growth	Basic (pH 11) Alkaline Lakes
High Temperature:	74°C Hot Springs		90°C Heat-Treated	112°C Dried
Low Temperature:	-269°C Liquid Helium	-196°C Liquid H$_2$	-55°C Frozen	-2 to +4°C Antarctic lakes
Desiccation:	88 yr Herbarium Specimen		103 yr Dried	Absence of Rainfall Atacama Desert
Oxygen:	<0.01% Anoxic Lakes	1% Blooms	20% Ambient O$_2$	100% Cultures
Carbon Dioxide:	0.001% Cultures	0.035% Ambient CO$_2$	3.5% Cultures	40% Cultures
Radiation:	Ultraviolet 290 to 400 nm*	X-Rays 200 kr†	γ-Rays 2,560 kr‡	Highly Ionizing Thermonuclear Bomb

*Absorbed by sheath. †Twice as resistant as single-celled algae. ‡Ten times as resistant as single-celled algae.

Figure 2.9. Growth and survival of cyanobacteria.

Why are cyanobateria so tolerant?

Life survives by fitting to its surroundings. But for organisms such as cyanobacteria, members of evolutionary branches that date from the very distant geological past, the surroundings changed as the global environment evolved over huge spans of time. Cyanobacteria adapted as the environment of the Earth evolved, but because they never lost their gene-encoded mastery of settings faced before, they developed enormous versatility and were themselves the root of the greatest environmental change ever to affect the planet – the onset of an oxygen-rich atmosphere.

Early in Earth history, when cyanobacteria first spread across the globe, free oxygen was in short supply. The oxygen their photosynthesis pumped into the surroundings was quickly scavenged by the environment, sponged up, for example, by combining with ferrous iron to form the iron oxide minerals sedimented in rocks known as banded iron formations. Because oxygen concentrations were low, there was no ultraviolet (UV)-absorbing ozone layer and the Earth's surface was bathed in a deadly stream of UV light. Cyanobacteria faced a quandary. They needed sunlight to power photosynthesis, but if they lived in shallow waters where light was strong, they would die, the overlying water too shallow to shield them from the lethal radiation. The earliest evolved members of the group countered this threat by living deep under water, using gas-filled cellular pockets (vesicles) to control their buoyancy and having photosynthetic machinery that operated in exceedingly faint light, a strategy used today by the cosmopolitan and exceptionally abundant marine cyanobacterium *Synechococcus*.

Over time, oxygen and ozone began to build up, but concentrations remained low as oxygen continued to be sponged up and UV remained a lethal threat. To colonize shallow-water settings, the best environments for their light-dependent photosynthetic lifestyle, cyanobacteria invented biochemical means to repair UV-caused cellular damage plus other protection mechanisms: colonial ball- and jelly-bean-shaped varieties (chroococcaceans and entophysalidaceans) ensured the cover of overlying waters by cementing themselves to the shallow seafloor with gelatinous mucilage that in some species was infused with a UV-absorbing biochemical, scytonemin; and bottom-dwelling string-like oscillatoriaceans, able to glide toward or away from light depending on its intensity, entwined themselves in felt-like mats that blanketed shallow basins. But probably most importantly, as a byproduct of their photosynthesis they spewed gaseous oxygen into their immediate surroundings, giving themselves a telling advantage over their oxygen-sensitive microbial competitors (non-oxygen producing photosynthetic bacteria) in the battle for photosynthetic space. In modern parlance, cyanobacteria were the first global polluters, pumping a noxious – in fact highly toxic – gas into the environment. But in evolutionary terms, in the fight for survivial and ultimate success, their 'gas warfare' won the day.

Thus, cyanobacteria triumphed wherever photosynthesis could occur, from the open ocean to nearshore shallows; in lagoons, lakes, seas, and streams; in frigid to blistering hot locales; from exposed mudflats to deserts and the rocky land surface. The survive-then-thrive-almost-anywhere lifestyle of these remarkable living fossils enabled them to take over the globe. Some experts claim that living fossils are simply champions at warding off extinction. If so, cyanobacteria must be, over all of geological time, the true Grand Champions!

The rise of eukaryotic (nucleated) life

In the formal classification of living systems, each organism belongs to one of three major superkingdom-like groups (known as 'domains'): (1) Bacteria, the domain composed of non-nucleated microorganisms such as cyanobacteria and all of the many different kinds of bacterial microbes; (2) Archaea, a second major group of non-nucleated microorganisms that includes methane-producing microbes and various 'extremophiles' that thrive in exceedingly acidic, high-temperature settings; and (3) Eucarya, organisms such as algae, fungi, plants, and animals that are made up of cells in which chromosomes are packaged in a prominent balloon-like sac, the cell nucleus. Rather than being enclosed in such a sac-like body, the genetic material (DNA) of Bacteria and Archaea is dispersed throughout their cells. Thus, whether large or small, living or fossil, the organisms that make up the three domains of life are of only two basic types: non-nucleated prokaryotes (from ancient Greek, 'before the nucleus'), members of the Bacterial and Archaeal domains; and nucleus-containing eukaryotes (from the Greek, 'truly nucleated'), members of the Eucarya such as ourselves.

The earliest eukaryotic organisms seem certain to have been small single cells that except for the presence of a cell nucleus would have closely resembled their prokaryotic ancestors. But because fragile intracellular bodies, such as nuclei, are virtually never preserved in fossils, we cannot expect the earliest eukaryotes to be easily identifiable in the fossil record. After such early eukaryotes had evolved and become somewhat diverse, however, the presence of eukaryotes would have been evidenced by their relatively large cells. For example, the cells of ball-shaped prokaryotes are tiny, almost all smaller than 5 μm; few are as large as 10 μm; and there are only two 'large' species, both less than 60 μm across. Cells of eukaryotes ordinarily are much larger – tens, hundreds, even thousands of microns across. So, single-celled fossils in the 10–60 μm range have been regarded as 'possible eukaryotes', and those larger as 'assured eukaryotes'. Judging from living organisms, the 60 μm boundary seems a safe limit; no prokaryotic unicells are larger and practically none come close.

Although hints of eukaryotes have been reported from rocks 2100 Ma in age and even older, the earliest undoubted evidence of the group dates from about 1800 Ma ago: simple balloon-like unicells are known from near the farm town of Jixian, east of Beijing, China, as well as from deposits of about the same age in Russia and the Ukraine. Because the Jixian and similar-aged unicells are large, some more than 200 μm across, there can be little doubt they are fossil eukaryotes. As their preserved alga-like cells are scattered across the remnants of an ancient seafloor, we know they are fossil algal phytoplankton. However, though unquestionably alga-like, their simple morphology – ball-shaped, but varying from smooth, to rough, to spiny (Figure 2.10) – provides too little information to indicate their exact relations to living algae (for which reason they are therefore formally referred to as acritarchs, members of the taxonomic group Acritarcha, from the Greek, *akritos*, 'confused, uncertain').

Eukaryotes perfect the art of cloning

Like some species of single-celled algae living today (*Chlorella*, *Chlorococcum*, and their relatives), early acritarchs reproduced by mitosis – i.e. 'body cell division'. This is the simplest way for a eukaryotic cell to multiply – a parent cell merely clones itself into two exact copies. Most prokaryotes reproduce by a similar process, cell fission, in which the single strand of genetic material in a parent cell is duplicated and the copies are passed to daughter cells formed as the parent splits in half. But cloning by mitosis in eukaryotes is more complicated. The cells of eukaryotes usually have many strands of DNA, packaged in chromosomes, and these are cordoned-off in a nuclear sac that must be broken down before the copied chromosomes can be passed to offspring. As shown in Figure 2.11, the chromosomes are duplicated; freed from the nucleus and aligned near the centre of the cell; then pulled apart into

52   J. William Schopf

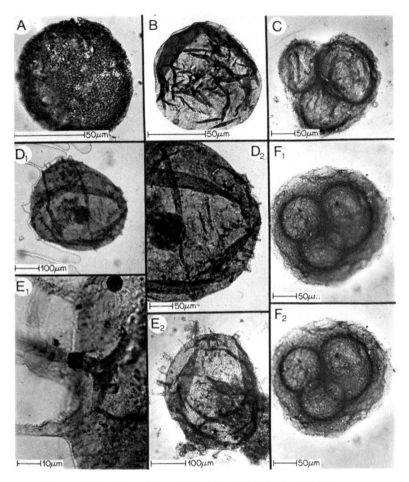

Figure 2.10. Flattened, originally spheroidal, single-celled algae
(acritarchs) from the 950-million-year-old Lakhanda Formation of Siberia.

two new cells, each a clone of the parent. Mitosis is a highly organized
yet simple way to make new cells that are faithful copies of the old.

In its immediate results, the mitotic (body cell division) form of
reproduction practised by these early eukaryotes can be thought of as
a eukaryotic version of prokaryotic fission – both produce offspring
that are unchanged copies of their dividing parents. So, it is not surprising that the long-term evolutionary results of the two processes
are also very similar – primitive algal acritarchs, like non-nucleated
cyanobacteria, evolved at an almost imperceptibly slow pace. Though
mitotic phytoplankton were an important stage-setter for later evolutionary advance, the early evolved, most primitive kinds were masters

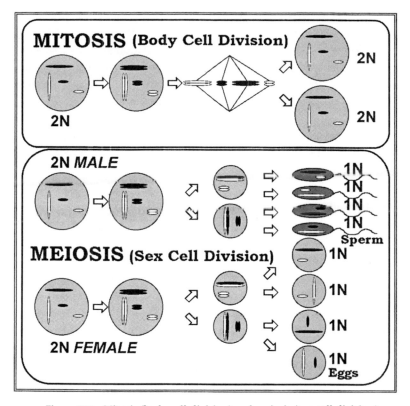

Figure 2.11. Mitosis (body cell division) and meiosis (sex cell division).

of status quo evolution, changing little over many hundreds of millions of years.

Sex: a key advance

Virtually all eukaryotic cells, whether present in microscopic single-celled algae or in large many-celled organisms such as ourselves, can divide by mitosis which not only makes new cells as an organism grows but replaces those that have become damaged or aged. But though mitosis can make new cells to replace the old – and in very simple animals (such as, for example, the tiny freshwater animal *Hydra*) and in some plants (such as those that multiply by means of subterranean 'runners') can generate whole new organisms – it is not the mechanism used to make offspring in advanced forms of life. Instead, in humans, most other animals, and the great majority of higher plants, reproduction takes place by the fusion of egg and sperm, specialized sex cells (gametes), that are made by the process of meiosis, an advanced type of cell division that halves the number of chromosomes. The original number is put together again at fertilization to give the fertilized egg, the zygote, a complete set of chromosomes, half coming from each parent. In animals, meiosis makes sex cells; in plants, meiosis makes spores which then divide later by mitosis to produce eggs and sperm.

Meiosis starts out much like mitotic cell division, its evolutionary predecessor. In the first step, the chromosomes double as they do in the first step of mitosis. But instead of being shunted to *two* new cells, as in mitosis, the paired chromosomes split once and then a second time and are distributed to *four* cells, each of which has only half the number of chromosomes of the starting cell (Figure 2.11).

The mitotically dividing body cells of eukaryotes such as fungi, plants, and animals contain two copies of each chromosome, a complement abbreviated as '2N' (where the N stands for number) and known technically as the diploid number. In humans, for example, each body cell contains 23 pairs of chromosomes, so the 2N (diploid) number is 46. In contrast, human sex cells, eggs and sperm formed by meiosis, contain only 23 chromosomes, a set that represents half that of each body cell and known as the '1N' or haploid number. During the life cycle of all advanced eukaryotic organisms, mitosis and meiosis both occur, each playing its particular role (Figure 2.12). Thus, for example, in animals the mitotically dividing cells of an adult are all 2N, each housing two copies of the chromosomes. Certain of these cells undergo changes that enable them to divide by meiosis to form eggs or sperm, 1N sex cells. When fertilization occurs, the 1N gametes fuse to form a 2N zygote. And once the zygote grows to an adult by mitosis, this sexual life cycle then repeats, with the formation of new sex cells, new zygotes, new adults, and so on.

Why does sex matter?

Among all inventions evolution ever devised, only two stand out as surpassingly important: (1) oxygen-generating (cyanobacterial) photosynthesis, key to the development of the oxygen-dependent workings of the modern living world; and (2) eukaryotic sex, the main source of genetic variation in higher organisms and the root of their diversity and rapid evolution.

The pre-sex living world was more or less static, with evolution being incredibly slow. From time to time, new well-equipped mutants

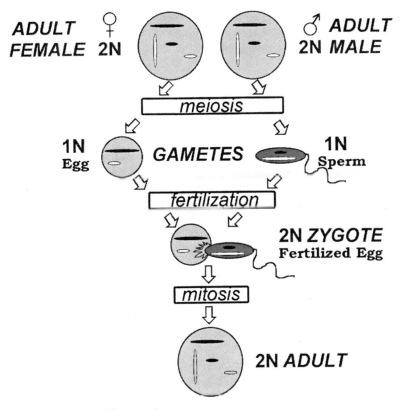

Figure 2.12. Life cycle of a sexually reproducing eukaryote.

emerged among the non-sexual microbes and mitotic single-celled algae. Yet this happened only rarely because most mutations are harmful (as useful ones are quite scarce) and reproduction by cloning maintained the genetic status quo. But everything changed when cloning was replaced by meiotic sex, a breakthrough that from the very start added a huge supply of grist to the evolutionary mill.

During meiosis, chromosomes often exchange parts to form new combinations of genes, so the suites of chromosomes parcelled out vary among the gametes. Because the odds of having exactly the same genes present in any two eggs or in any two sperm are infinitesimally small – and the odds of having the precise same set of genes in any two fertilized eggs are even smaller still – every organism born from sexual reproduction will contain a genetic mix that has never existed before. Even children of the same family always have somewhat different combinations of genes (except for identical twins, developed from a single fertilized egg). Of course, all of us have many genes in common with our sisters, brothers, and parents (and for this reason share with them what we call family resemblance), but we are not photocopies, not clones like Dolly the sheep.

Sex brings specialization

Because the mix of genes in each individual is unique, a population of sexual organisms includes an enormous number of different gene combinations (a situation that constrasts sharply with the monotonously uniform genetic make-up of earlier evolved, non-sexual species). This huge diversity, a direct result of sexual reproduction, explains why the advent of sex was such a breakthrough. By forming countless new mixtures of genes, sex rapidly increased the amount of variation within species. This led to increased competition among members of each species – for example, for food resources or for the best mates – and this, in turn, spurred development of even more novel gene combinations and the emergence of new biological species. So, the advent of sex had a telling impact on the history of life, not only by increasing the range of capabilities of already established species but by speeding the genesis of brand new kinds of organisms.

But sex brought other changes as well. Most important was the division of the life cycle into two separate parts: one devoted to an organism's mitotic growth in size; and the other, to its meiotic-based reproduction. Over time, each part became more and more adept at its particular task and by doing so gave rise to an ever-increasing diversity of different types of eukaryotes – some specialized for growth, others for reproduction, still others proficient at both, and all especially well-suited to fit their own particular niche in the environment. The prominence of sex in the Phanerozoic and its absence throughout most of the Precambrian decisively divides these two great epochs in the history of life.

EVOLUTION EVOLVED

The advent of sex in eukaryotes, evidently about 1000 Ma ago, was thus a pivotal point in the evolution of evolution. Sex not only increased variation within species, diversity among species, and the speed of evolution and genesis of new species, but it also brought the rise of organisms specialized for particular habitats. By early in the Phanerozoic, the adult (mitotic) part of life cycles in plants had come to focus on vegetative growth, giving rise as the eon unfolded to trees, shrubs, and grassy vegetation honed to specific settings. At the same time, the meiotic, sexual part of plant life cycles came to be specialized for increasingly reliable reproduction as the flora evolved from spore-producers to wind-pollinated seed plants (gymnosperms) and, eventually, to flowering plants (angiosperms) where propagation is often aided by pollen-carrying insects. Animal evolution followed a parallel course: in adults (the mitotic part of the life cycle), changes in limbs and teeth led to increasingly effective, specialized ways of hunting, foraging, and feeding, while the meiotic part of the life cycle evolved improved means of reproduction – from the broadcast larvae of marine invertebrates and roe of fish to the soft egg masses of amphibians, the hard-shelled eggs of reptiles and birds, and the protected embryos of mammals.

Phanerozoic plants and animals mostly are large; all are many-celled; because of their sexual life cycles, are typically specialized for particular settings; and, generally, were relatively short-lived, existing on average for some 10 or so million years. As a result, the Phanerozoic was punctuated by repeated episodes of extinction (as we will see in the other chapters of this book), often followed by the rise and diversification (adaptive radiation) of the lineages that survived. But the earlier and much longer Precambrian was vastly different. Almost all species were microscopic, non-sexual, and prokaryotic. For the most part, evolutionary changes seem to have occurred in their biochemical make-up, not in their overall morphologies. The most successful Precambrian microorganisms were cyanobacteria, exceedingly long-lived generalists able to tolerate a wide range of environments and, thus, to survive essentially unchanged over incredibly long segments of geological time. And decimating mass extinctions, like those that from time to time cut great swaths though the Phanerozoic biota of eukaryotic specialists, simply did not occur among these earlier-evolved prokaryotes.

The take-home lesson is clear (Figure 2.13): life's history is divided into two great epochs, each having its own biology, style, and tempo – the Precambrian 'Age of Microscopic Life', a world of microbial long-lived ecologic generalists; and the Phanerozoic 'Age of Evident Life', ruled by eukaryotic short-lived specialists. Evolution, itself, evolved!

If it's not broken, don't fix it

At first glance, the notion that the very rules for the survival and success of life changed – that from the Precambrian to the Phanerozoic,

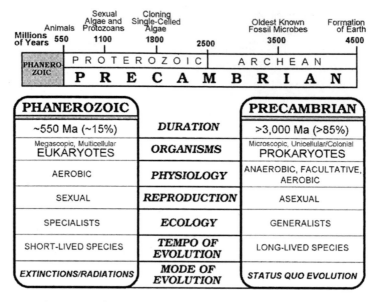

Figure 2.13. The rules of evolution changed from the Precambrian to the Phanerozoic. Ma, million years.

evolution, itself, evolved – may seem a bit hard to swallow. Indeed, this notion fits not at all our mind's eye image of life's history as a steadily changing parade from small to large, simple to complex, a gradually unfolding sequence in which evolution and change go hand-in-hand. Yet, for much of the earliest 85 per cent of the history of life, this commonly taught view does not fit the facts, and the explanation turns out not to be surprising. In actuality, the aim of life is never to evolve, never to change at all! Life's true slogan might well be 'if it's not broken, don't fix it', for when systems do break down, for instance by mutation, living systems respond immediately by repairing the mutant part back to what it was before. We see evolutionary change only because life's history spans such an exceedingly long time that rare unfixed mutations add up.

So, rather than change, life's goal is actually to maintain the status quo. In this light it seems not at all unexpected that simple prokaryotes, such as cyanobacteria, and the earliest evolved eukaryotes, mitotic single-celled algae that mutiplied by cloning, changed little or not at all over hundreds of millions of years. This strategy of almost imperceptibly sluggish evolutionary change was successful for literally billions of years.

Life's two-part history

To grasp better how the two-part, Precambrian-to-Phanerozoic history of life fits together, let us consider a simple analogy drawn from American history. The similarities are of course rough, but it is interesting to see how settlement of the North American frontier by the early pioneers appears to parallel the palaeontological past.

As America came into existence, the first wave of settlers streaming west included few true craftsmen, few settlers specialized in the building of houses, the making of crockery or the weaving of clothes. But needs then were the same as now – a place to live, plates to eat from, clothes to wear. So, like generalists of the Precambrian, the most successful pioneers were jacks-of-all-trades, masters of survival who could cope with almost any situation. As settlements grew, so did specialization, with the arrival of merchants, tradesmen, and artisans. A new breed of house-builders could build, but not make plates or clothes. Clothiers could clothe, but not erect a house. Products were made better, faster, and could be sold at far less cost. And as in the history of Phanerozoic life, specialization led to diversity as crafts spawned more and more new subcrafts – roofers and plumbers to assist the house-builders, button-makers to supply the clothiers. But also as in the Phanerozoic, such specialization upped the odds of extinction. Button-makers lost out when zippers were invented; candle-makers, when gas lamps became the vogue; gas lamp-lighters, when electricity came to be commonplace; blacksmiths, when automobiles supplanted horse-drawn carriages. Specialization and short-lived success go hand-in-hand. Diversity abounds, but the varied components are ever-changing.

Specialization also breeds interdependence and this, too, spurs extinction. Even today the rise of computers has not only put type-writer manufacturers out of business but people dependent on them as well – for example, typewriter repair teams and suppliers of parts. The biology of ecosystems is interdependent in much the same way, so it is not surprising that a similar rippling of cause-and-effect happened many times during the Phanerozoic when extinction wiped out key components at the base of the food web.

As in the Phanerozoic, specialists predominate in our modern citified existence whereas jacks-of-all-trades – a local handyman, for instance – are hard to find. We accept this as 'progress' and, of course, it is. But such specialization also has its downside. The next time your computer crashes, or your cellphone is out of order, or your car won't start, remember the early settlers; like generalist cyanobacteria of the Precambrian, they would be able to cope!

ACKNOWLEDGMENTS

This essay is based on material developed at greater length in *Cradle of Life, The Discovery of Earth's Earliest Fossils* (J. W. Schopf, 1999, Princeton University Press), and draws particularly from chapters 1, 3, 8, 9, and 10 therein. I thank J. Shen-Miller for helpful suggestions, and Paul D. Taylor, the editor of this volume, for his friendship and patience.

FURTHER READING

Cloud, P. A working model for the primitive Earth. *American Journal of Science* **272** (1972), 537–48.

Beginnings of biospheric evolution and their biogeochemical consequences. *Paleobiology* **2** (1976), 351–87.

Golubic, S. Organisms that build stromatolites. In: Walter, M. R. (Ed.), *Stromatolites, Developments in Sedimentology 20*. Amsterdam: Elsevier, 1976, pp. 113–26.

Knoll, A. H. and Bauld, J. The evolution and ecologic tolerance of prokaryotes. *Transactions of the Royal Society of Edinburgh: Earth Science* **80** (1989), 209–23.

Knoll, A. H. and Golubic, S. Proterozoic and living cyanobacteria. In: Schidlowski, M., Golubic, S., Kimberley, M. M., McKirdy, M. M. and Trudinger, P. A. (Eds.), *Early Organic Evolution*. New York: Springer, 1992, pp. 450–62.

Schopf, J. W. Times of origin and earliest evidence of major biologic groups. In: Schopf, J. W. and Klein, C. (Eds.), *The Proterozoic Biosphere, A Multidisciplinary Study*. New York: Cambridge University Press, 1992, pp. 587–91.

Disparate rates, differing fates: Tempo and mode of evolution changed from the Precambrian to the Phanerozoic. *Proceedings of the National Academy of Sciences USA* **91** (1994), 6735–42.

Cradle of Life, The Discovery of Earth's Earliest Fossils. Princeton: Princeton University Press, 1999.

The paleobiologic record of cyanobacterial evolution. In: Brun, Y. V. and Shimkets, L. J. (Eds.), *Prokaryotic Development*. Washington, DC: American Society of Microbiology, 2000, pp. 105–229.

Simpson, G. G. *Tempo and Mode in Evolution*. New York: Columbia University Press, 1994.

Stanley, S. M. *Extinction*. New York: Scientific American Books, 1987.

REFERENCES

Green, J. W., Knoll, A. H. and Swett, K., 1988. Microfossils from oolites and pisolites of the Upper Proterozoic Eleonore Bay Group, central East Greenland. *Journal of Paleontology* **62**: 835–852.

SCOTT L. WING

Department of Paleobiology, Smithsonian Institution, Washington, DC, USA

3

Mass extinctions in plant evolution

INTRODUCTION

Mass extinctions generally are recognized as major features in the history of life. They sweep aside diverse, sometimes even dominant groups of organisms, freeing up resources that can then fuel the diversification and rise to dominance of lineages that survived the mass extinction. This view of the history of life has been developed in large part from the study of shelly marine animals, and to a lesser extent from studies of terrestrial vertebrates (e.g. Valentine, 1985). By compiling data on the stratigraphic ranges of genera and families of marine animals, palaeontologists have been able to recognize the 'Big Five' mass extinctions, occurring at the end of the Ordovician, in the Late Devonian and at the end of the Permian, Triassic and Cretaceous periods (e.g. Sepkoski, 1993; Chapters 1 and 5). Each of these episodes is a geologically sudden decrease in taxonomic diversity. Terrestrial vertebrates also show major declines in taxonomic diversity at the end of the Permian and at the end of the Cretaceous (Benton, 1993). In contrast, compilations of the stratigraphic ranges of species of land plants do not show major declines in diversity (Niklas *et al.*, 1980, 1985; Niklas and Tiffney, 1994; Figure 3.1). The absence of major declines in the diversity of land plants as represented in these compilations of stratigraphic ranges has led to the suggestion that plants are more resistant to mass extinctions than animals (Niklas *et al.*, 1980; Knoll, 1984; Traverse, 1988).

Knoll (1984) identified three aspects of the biology of plants that might tend to make them resistant to some causes of mass extinctions, such as the impact of a large asteroid, but vulnerable to other factors like climate change. First, most plants are quite resistant to physical

Extinctions in the History of Life, ed. Paul D. Taylor.
Published by Cambridge University Press. © Cambridge University Press 2004.

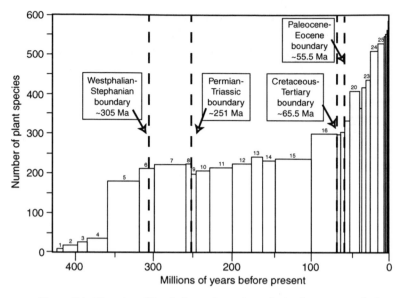

Figure 3.1. Diversity of land plants through geologic time as compiled
from the temporal ranges of 8688 species by Niklas and Tiffney (1994).
Note there are no major declines in species diversity. Among marine
animals the two largest mass extinctions are at the Permian–Triassic
(estimated 80 per cent species extinction) and Cretaceous–Tertiary
boundaries (estimated 60 per cent). Among plants there is about a
10 per cent decrease in species diversity at the Permian–Triassic
boundary, and almost no decrease in diversity at the Cretaceous–Tertiary
boundary. Each bar represents an interval of geological time: 1, Late
Silurian; 2, Early Devonian; 3, Middle Devonian; 4, Late Devonian;
5, Mississippian (Early Carboniferous); 6, Pennsylvanian (Late
Carboniferous); 7, Early Permian; 8, Late Permian; 9, Early Triassic;
10, Middle Triassic; 11, Late Triassic; 12, Early Jurassic; 13, Middle
Jurassic; 14, Late Jurassic; 15, Early Cretaceous; 16, Late Cretaceous;
17, early Paleocene; 18, late Paleocene; 19, early Eocene; 20, middle
Eocene; 21, late Eocene; 22, early Oligocene; 23, late Oligocene; 24, early
Miocene; 25, late Miocene; 26, early Pliocene; 27, late Pliocene; 28, early
Pleistocene; 29, late Pleistocene. The width of each bar is proportional to
the duration of the interval as given in the 2003 time scale of the
International Commission on Stratigraphy. Redrawn from figure 1 of
Niklas and Tiffney (1994). Intervals discussed in detail are indicated by
dashed lines. Ma, million years.

destruction. They can regrow limbs that are damaged, resprout from roots or stumps even if the whole above-ground plant is destroyed, and they produce seeds or spores that can reside in the soil during unfavourable periods only to sprout and grow at a later date. In extreme examples seeds may persist for more than 1000 years and remain viable (Shen-Miller et al., 1995).

A second aspect of plant biology that might create a different pattern of extinction than seen in animals is that plants are sessile, that is, they do not move as adults, though of course the geographic distributions of plant populations change through dispersal of seeds. In order to grow and mature in a given location, a plant has to be able to survive changing conditions or complete its life cycle in between fatal events. If the environmental tolerances of a plant are exceeded there is no escape, which puts a premium on the ability to tolerate change or to escape bad conditions through dispersing seeds or spores. As a result, many plants can disperse long distances, and populations of plants have been documented to change their distributions at rates averaging more than a kilometre per year (Clark et al., 1998).

A third aspect of plants that might affect their extinction response is the way in which they obtain food. Almost all plants (except for some parasites and saprophytes) make their own food from sunlight, water and atmospheric CO_2. In contrast, animals are differentiated by trophic levels: there are herbivores (primary consumers), secondary consumers that eat herbivores, tertiary consumers that eat secondary consumers, etc. Furthermore, animals are differentiated from one another by moving in different ways to obtain their food, and/or by having their active periods at different times of day. Knoll (1984) hypothesized that larger overlap in resource requirements among plants, combined with the wide (though of course not uniform) availability of those resources, could lead plants to compete more strongly with one another than do animals. Stronger competition among plant species than among animal species might make extinction from competition relatively more important in plants.

If mass extinction is relatively unimportant among plants and 'background' extinction resulting from competition is relatively important, this should have an effect on the geological history of plant diversity (Knoll, 1984; Niklas et al., 1985). We would expect few or no large and rapid reductions in plant diversity associated with impact events (because of the physical resilience and dormancy mechanisms of plants). We would also expect that the evolution and radiation of new lineages might accelerate extinction rates among pre-existing ones. The plant fossil record might be expected to show many examples of gradual competitive replacement of one lineage by another, and few examples of sudden elimination of lineages by devastation. This is consistent with the general patterns of diversity change observed in large-scale compilations of stratigraphic ranges (Niklas et al., 1985; Niklas and Tiffney, 1994; Figure 3.1).

Although mass extinctions have not been observed in large-scale compilations of stratigraphic ranges of plants, there are nine intervals during which extinction rate rose above background levels: two intervals during the Early Devonian, and one each during the Late Devonian, Late Pennsylvanian–Early Permian, Early Triassic, Late Jurassic, Early Cretaceous, Oligocene, and Miocene (Niklas, 1997). None of these intervals with elevated extinction rates coincide with the major mass extinctions identified for marine or terrestrial animals, and only two, the Pennsylvanian–Permian and the Early Triassic coincide with any overall decline in plant diversity (Niklas and Tiffney, 1994; Figure 3.1). The increase in extinction rate without a corresponding decrease in overall diversity implies that during most of these intervals there must have been compensating increases in speciation rate among some lineages of plants. There is evidence that these increases in extinction rate affected some lineages more than others. For example, during the Devonian gymnosperms (seed plants) had higher extinction rates than ferns and their relatives, whereas during the Early Permian the reverse was true (Niklas, 1997). Niklas (1997) suggested that these differences among lineages in extinction rates reflect differences in their reproductive biology, with, for example, gymnosperms being less severely affected by climatic drying during the Early Permian than ferns and their allies because the latter require water for reproduction (see Appendix 3.1). Figure 3.2 shows the proposed relationships between the major groups of land plants.

All of the work mentioned above relies on compilations of the published stratigraphic ranges of plant species. The geographic scope of the compilations is nominally global, but favours North America and Europe because of the long history of palaeobotanical study on these continents. It is not known if patterns of diversity-change over time differ between continents. The temporal resolution of these compilations is also fairly coarse, with intervals from 5 to 10 or more million years long being used in describing stratigraphic ranges (Niklas et al., 1980; Niklas and Tiffney, 1994). Although the idea has not been tested directly, it is possible that plant extinctions of short duration or sub-global geographic scope have been missed because of the

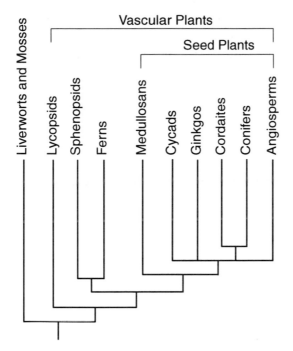

Figure 3.2. A phylogenetic diagram showing proposed relationships between major groups of land plants based on the work of Kenrick and Crane (1997). See Text Box 3.1 for descriptions of major groups of plants.

coarseness of the time scale and the analysis of data from multiple continents.

Below I examine four intervals in the history of terrestrial plants for which diversity and species composition has been studied on a much finer scale. The studies are regional rather than global in geographic scope, use the finest level of taxonomic sorting that can be achieved, and divide time as finely as local stratigraphic sections and sampling allow – typically into intervals of 10^4–10^5 years. The detailed studies reveal regional changes in plant diversity that would be missed in a broader study, but may not be typical of a global pattern (if indeed there is one global pattern). The four episodes examined in this chapter are the Westphalian–Stephanian boundary (*c.* 305 million years (Ma) ago), the Permian–Triassic boundary (*c.* 251 Ma ago), the Cretaceous–Tertiary boundary (*c.* 65.5 Ma ago), and the Paleocene–Eocene boundary (*c.* 55.5 Ma ago). These intervals, shown in Figure 3.1, were chosen because detailed studies of floral change have been carried out in at least some regions. I will characterize each episode in terms of how severely land plants were affected, which plant groups were hit hardest, how quickly the extinction occurred, how large an area was affected, whether there were concurrent extinctions among animals and what sorts of environmental changes might have caused the extinction. As I examine individual extinctions and compare them to one another I will consider how the biological features of plants listed above might affect their responses to the stresses that cause mass extinctions. After the review I will assess whether plants do indeed show a different pattern of mass extinctions than animals, and explore the implications of these differences for the causes of mass extinctions.

CASE STUDIES OF PLANT EXTINCTIONS

The Westphalian–Stephanian extinction (c. 305 Ma ago)

The Carboniferous and Early Permian were the last time, prior to the most recent few million years, that the Earth experienced major continental glaciation outside the polar regions. The Carboniferous glaciation was confined mostly to the southern supercontinent of Gondwana, but it had global effects on climate and sea level (Rowley et al., 1985). The cold polar regions of the time are thought to have restricted the wet tropical belt to a fairly narrow range of latitudes that included most of what is now eastern and southern North America and western Europe – a region called tropical Euramerica. Tropical Euramerica had vast coastal plains where the ocean came and went in synchrony with the waning and waxing of glaciers in Gondwana. Extensive swamps formed on the tropical coastal plains, especially during times when the ocean retreated, and plant matter built up in the swamps because it decayed slowly in the wet conditions. These accumulations of plant matter, originally peat, are now the major coal deposits of eastern North America and western Europe.

Although this is the oldest of the plant extinctions to be considered here, it is in some ways better known than any of the others. Many of the fossils that document plant life in the Carboniferous come from coal balls, which are masses of peat that were infiltrated with calcium carbonate before they were compressed. This early mineralization preserved the plants in cellular detail, which improves confidence in their identification. Furthermore, coal balls have been used to estimate the relative amount of biomass produced by each species within an ancient peat swamp. (See Appendix 3.2 for a summary of plant fossil preservation and how this can affect the detection of extinctions.)

During the Westphalian Stage of the Late Carboniferous peat swamp vegetation was dominated by plants belonging to groups that are now extinct or much reduced in diversity (see Appendix 3.1 for brief descriptions of some of these plants). Some 40–50 species of plants are typically found in a single coal bed (DiMichele and Phillips, 1996). The major trees in the peat swamps were lycopsids. These plants were particularly well suited to the wettest parts of the landscape and their remains form more than half of the bulk of most coals of this age. Slightly drier areas were inhabited by a variety of small ferns and tree ferns of the family Marattiaceae. Fossils of these plants are associated with fossil charcoal, indicating frequent fires. Other plants common in lowland swamps were seed ferns like *Medullosa*, and conifer-relatives called cordaites. Tree-sized horsetails (calamites) were also moderately abundant, though probably more common in muddy swamps than in peat swamps proper.

The extinction at the Westphalian–Stephanian boundary (*c*. 305 Ma ago) is thought to have occurred over an interval of 100 000 years or less. Data from coal seams in eastern North America show that at this time about 67 per cent of the species inhabiting the peat swamps were eliminated, and some 50 per cent of the species found in muddy swamps (DiMichele and Phillips, 1996; Figure 3.3). Pollen and spores show approximately 44 per cent extinction across the same time interval (DiMichele and Phillips, 1996). The extinction did not hit all forms equally. Of the tree species in the peat swamps (mostly lycopsids and tree ferns) almost 87 per cent did not survive to the Stephanian. Among smaller plants (ground cover, vines and shrubs) the extinction was only about 33 per cent. The lycopsids were hit hard not only in terms of species, but also in terms of abundance. Coal balls from the late Westphalian are 60–70 per cent lycopsids by volume, but by the early Stephanian they were less than 10 per cent (DiMichele and Phillips, 1996).

Peat swamps continued to develop in Euramerica during the Stephanian. Each of the first few coals to form after the extinction event has different species composition and dominants, but after that there is a consistent pattern of domination by tree ferns rather than by lycopsids. In general the Stephanian peat swamps appear to have been more homogeneous habitats, with less area of the standing water environment that had been the habitat most dominated by lycopsids (DiMichele and Phillips, 1996).

The extinction was geologically rapid and has been documented through much of tropical Euramerica; however, different species

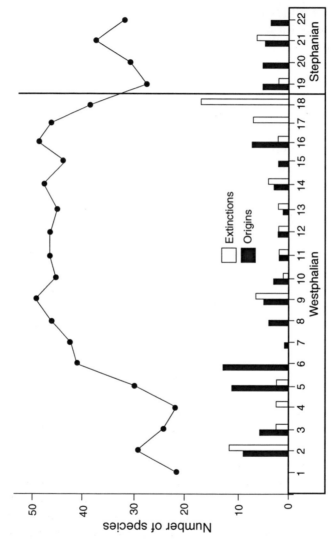

Figure 3.3. Record of plant extinctions at the Westphalian–Stephanian boundary (modified from DiMichele and Phillips, 1996). Each of the numbered intervals represents a coal bed from which fossils were studied. Note the high level of extinction and low origination in the last Westphalian coal.

belonging to the same higher taxa (genera and families) continued to form lycopsid-dominated peat swamp floras in parts of China through the rest of the Late Carboniferous and into the Early Permian. Most authors attribute the reduction and eventual extinction of lycopsid-dominated peat swamps to the drying of the Euramerican tropics as the continents moved northward and glaciations in Gondwana declined (Kerp, 2000).

The dominance of ferns in the aftermath of the Westphalian–Stephanian extinction has been said to reflect the success of plants with high dispersal ability, rapid growth and relatively broad environmental tolerances (DiMichele and Phillips, 1996). In other words, the extinction brought about a period of dominance by weeds.

The Permian–Triassic extinction (c. 251 Ma ago)

Traditionally palaeobotanists have recognized a transition between Pale-ophytic and Mesophytic floras that coincided roughly with the Early-Late Permian boundary (Kerp, 2000). Early Permian floras retained more marattialean tree ferns, seed ferns related to Medullosales, and in some areas even lycopsids, whereas Late Permian floras typically had more species of conifers, ginkgos, cycads and other seed plants. The grad-ual nature of this transition was emphasized by Frederiksen (1972) and Knoll (1984). Detailed studies of Early Permian plant fossils in Texas subsequently showed that the appearance of plants characteristic of the Mesophytic flora is associated with the drying out of the lowland basins where plant fossils are preserved (DiMichele and Aronson, 1992).

Another major change in floral composition has long been rec-ognized to have occurred near the Permian–Triassic boundary in the southern supercontinent of Gondwana (Retallack, 1995). In Australia, South America, Africa and Antarctica, most Late Permian floras are dominated in both number of species and number of specimens by a seed fern called *Glossopteris*. There were many species of *Glossopteris*, at least some of which were large trees with dense, conifer-like wood. *Glossopteris* had simple elliptical to obovate leaves with fern-like venation (hence the derivation of the genus name, which means 'tongue fern' in Greek), but the leaves had stalked seeds attached to them. Early Triassic floras in the same region are dominated by a different, distantly-related seed fern called *Dicroidium*, which had more complex, fern-like leaves. Many *Dicroidium* species appear to have been shrubby plants.

Studies of floral change and extinction near the Permian–Triassic boundary in Australia have cited up to 97 per cent extinction of plants

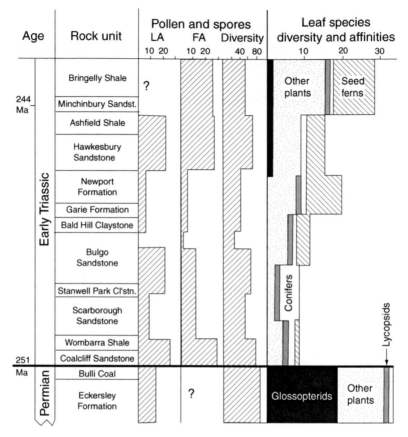

Figure 3.4. Record of plant extinctions at the Permian–Triassic boundary in eastern Australia based on megafossils and microfossils (modified from Retallack, 1995). The glossopterid seed ferns were most severely affected by the mass extinction at the end of the Permian in this area. Ma, million years.

(based on leaf fossils) in eastern Australian sequences (Figure 3.4; Retallack, 1995). Extinction levels in the microflora are much lower, about 19 per cent, perhaps partly because of reworking of Permian pollen into lowermost Triassic sediments (Retallack, 1995), but also because of the lower taxonomic resolution of microfloras.

It has become clear that there were dramatic changes worldwide in floras at the end of the Permian. A global survey of microfloras from about the last million years of the Permian shows that in many different local environments (shallow marine, various types of terrestrial setting) and on all continents there was a dramatic increase in the abundance

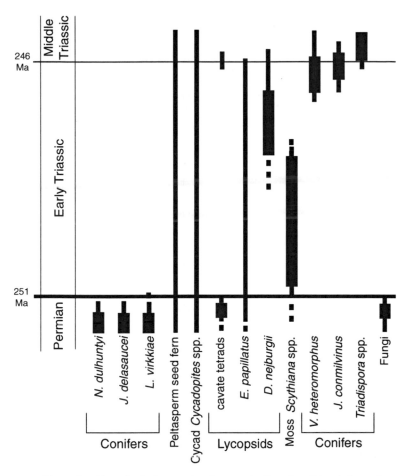

Figure 3.5. Record of plant extinctions at the Permian–Triassic boundary in Europe based on microfossils (modified from Looy et al., 1999). In tropical Euramerica conifers were most severely affected by the extinction (left side of diagram). Some seed ferns and cycads persisted, but the Early Triassic flora was heavily dominated by lycopsids (mostly herbaceous) and mosses. The conifers that appear at the end of the Early Triassic belong to different groups than the Permian conifers.
Ma, million years.

of fungal spores (Visscher et al., 1996). The global increase in fungal spores is a unique event in the history of life, and is thought to reflect a period during which there was massive die-off of terrestrial plants followed by fungal decay. The end of the Permian in the Euramerican region is now also known to be associated with the disappearance of many types of coniferous pollen at almost precisely the same time as

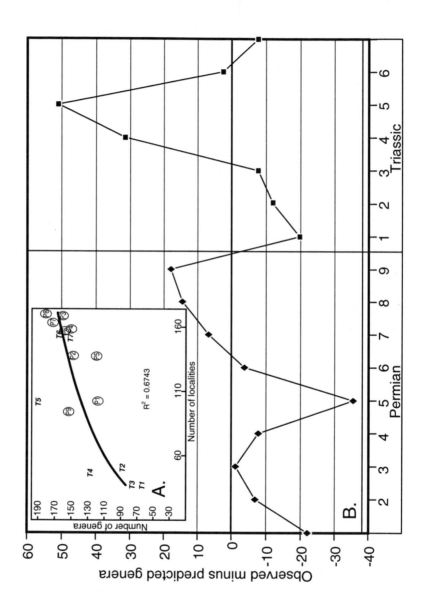

the extinctions in the marine realm (Looy et al., 1999; Twitchett et al., 2001; Figure 3.5). The cause of the die-off is uncertain, but it has been related to changes in the atmosphere resulting from the eruption of the vast Siberian flood basalts (Visscher et al., 1996; see also Chapter 5).

In spite of the strong pattern of plant extinction shown in many local sections, a global analysis of plant megafossils shows that the severity of extinction is highly variable in different areas (Rees, 2002). Globally about 60 per cent of plant genera were lost between the last stage of the Permian and the first stage of the Triassic, but the extinction also coincides with a steep decline in the number of localities from which fossil plants have been recovered (Rees, 2002). The ability to detect biological diversity is strongly determined by the number and size of samples, so with fewer fossil plant localities in the Early Triassic, it is not surprising that fewer species have been collected. This effect of sampling intensity is difficult to factor out, but by comparing the number of genera found in each stage of the Permian and Triassic with the number of samples from the stage it is possible to develop an expectation for how many genera will be found given a particular number of samples (Figure 3.6A). The earliest Triassic has few genera of plants even given the small number of samples (Figure 3.6B), so there may indeed have been a global decrease in generic diversity of plants, though the severity probably varied substantially by region (Rees, 2002).

Whatever the global effect on plant diversity, the collapse of terrestrial productivity in the latest Permian appears to have persisted in some areas for the first five to six million years of the Triassic. Worldwide there seems to have been no deposition of peat (coal) during the Early Triassic – the only period of geological time after the Devonian for which this is true (Retallack et al., 1996; see also Chapter 4). In tropical

←───

Figure 3.6. The effect of sample number on diversity in Permian–Triassic floras based on data published by Rees (2002). A (inset). Plot of the stages of the Permian (labelled P1–P9) and Triassic (T1–T7) comparing the number of genera in each interval with the number of samples. The regression line is the expected number of genera for a given number of samples. B (main graph). The difference between observed and expected (based on the regression in A) number of genera for each stage of the Permian and Triassic. Note there is a marked decline across the Permian–Triassic boundary, although two stages in the early and middle Permian have even fewer genera per sample (i.e. lower diversity) than does the first stage of the Triassic.

areas of what is now Europe microfloras show that the dominant conifers disappeared from the record at the end of the Permian; Early Triassic microfloras were composed largely of lycopsid and moss spores (Looy et al., 1999). The moss/lycopsid flora persisted for about five million years before conifer pollen once again became abundant. The long delayed recovery of terrestrial (and marine) ecosystems following the Permian–Triassic extinction is one of the most interesting and puzzling patterns observed for this time interval. It is still unclear if the length of the recovery interval is related to the severity of the extinction, reflecting the slow rediversification of life, or if it reflects continuing environmental stress of some sort (see Chapter 4). The underlying cause(s) of the Permian–Triassic extinctions (in both marine and terrestrial systems) are still poorly understood, but include rapid climate change resulting from increases in greenhouse gas concentrations, extraterrestrial impacts and oceanic anoxia (Erwin et al., 2002; Chapter 5).

The Cretaceous–Tertiary extinction (c. 65.5 Ma)

The mass extinction at the end of the Cretaceous Period is well known as the extinction that eliminated the dinosaurs, and is also the one mass extinction solidly connected with the impact of a large extraterrestrial object (Alvarez et al., 1980). Over the years since the Cretaceous–Tertiary (K–T) mass extinction was first linked to an impact, physical and chemical evidence for the impact has mounted. It is generally accepted that the object hit the earth at the Chicxulub impact site at the northern tip of the Yucatan Peninsula, leaving behind a slightly asymmetrical crater c. 150 km in diameter that implies an oblique angle of impact directed to the northwest (Hildebrand et al., 1991; Schultz and D'Hondt, 1996; see also Chapter 5).

The immediate results of the impact are reasonably well understood. A catastrophic blast, perhaps equal to c. 60 trillion tonnes of high explosives, occurred, and a high-speed, high-temperature shockwave was unleashed downrange toward North America (Schultz and D'Hondt, 1996). The impact also triggered massive slumping and debris flows on continental shelves (Norris and Firth, 2002). As much as 90 000 km^3 of target rock and bolide material were ejected from the impact site and distributed globally on ballistic trajectories (Claeys et al., 2002). Debris re-entering the Earth's atmosphere was heated by friction with the atmosphere, and started wildfires around the globe (Wolbach et al., 1988). Ejecta from the crater also included fine dust and minerals vaporized from the target rock and ocean water (Claeys et al., 2002). Because of the limestone and anhydrite rock at the impact site, the vaporized minerals would have produced CO_2, SO_2, and H_2O in the atmosphere (Pope, 2002).

Longer-term physical, chemical and biological effects of the bolide impact are less well understood. Initially Alvarez et al. (1980) calculated that the fine dust in the ejecta would have achieved a global distribution, and would have been so dense that there would have been insufficient light at the Earth's surface for photosynthesis for an extended period of months or years. It was inferred that temperatures would have plummeted because sunlight was blocked from reaching the surface (Pollack et al., 1983). This is referred to as the 'impact winter' hypothesis. Later estimates decreased the length of the impact winter to three to four months, and the most recent estimates of the quantity of fine dust produced by the impact suggest there may not have been enough in the atmosphere to diminish light levels below those required for photosynthesis (Pope, 2002).

Other authors have suggested that following the impact winter the CO_2 and water vapour created in the impact would have generated a global greenhouse effect that could have persisted for hundreds or thousands of years (O'Keefe and Ahrens, 1989; Wolfe, 1990). The CO_2 eventually would have been removed from the atmosphere by terrestrial and marine productivity, and through chemical weathering processes (Lomax et al., 2000). Another possible atmospheric effect is that generation of SO_2 and H_2O in the impact might have resulted in acid rain, and/or in global cooling as the result of SO_2 increasing the amount of sunlight reflected by the atmosphere (Pope et al., 1994; Pope, 2002). The fossil record of plants has played an important role in testing possible extinction factors at the K–T boundary.

The vast majority of detailed studies of plants across the K–T boundary have been carried out in western North America, and these studies have relied on all three types of plant fossils (Appendix 3.2). Studies of megafossils have revealed a very high level of species extinction, roughly 80 per cent, from the latest Cretaceous to the earliest Paleocene in western North Dakota (Figure 3.7; Johnson, 1992, 2002; Johnson and Hickey, 1990). This level of extinction has a high reliability because it is based on large, closely spaced samples that come from matched depositional environments and can be correlated with one another confidently through physical stratigraphy. High levels of extinction have also been recorded in megafloras from the Raton Basin of northern New Mexico and southern Colorado – about 1000 km closer to the Chicxulub impact structure (Wolfe and Upchurch, 1986). Levels of palynomorph

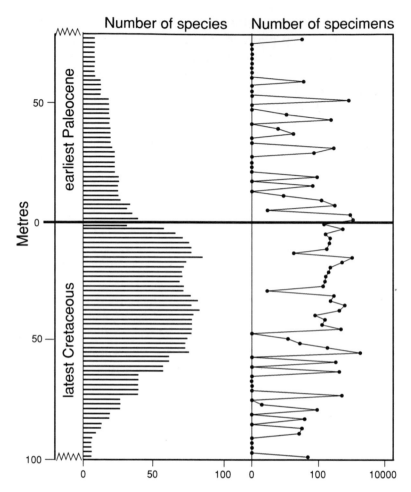

Figure 3.7. Record of plant extinctions at the Cretaceous–Tertiary
boundary in western North Dakota based on megafossils (modified from
Johnson, 2002). This record is based on fossil leaves. Note that the severe
decline in the number of species at the Cretaceous–Tertiary boundary is
not related to a decrease in the number of specimens collected, added
evidence that the extinction was a real event and not just a side effect
of irregular collecting effort.

extinction throughout the western interior of North America are
15–30 per cent (Tschudy et al., 1984; Johnson et al., 1989; Nichols and
Fleming, 1990; Nichols and Johnson, 2002; Hotton, 2002). As with the
Permian–Triassic extinction the difference in extinction rates between
megafloras and microfloras probably reflects the lower taxonomic resolution of palynomorphs (Nichols and Johnson, 2002; Appendix 3.2).

Studies of palynomorphs from the K–T boundary interval have revealed that the plant extinction is essentially coincident with the deposition of the impact ejecta from the K–T bolide (Hotton, 2002). Palynofloras demonstrate not only extinctions, but also vegetational disruption. Samples taken from the first few centimetres of rock overlying the impact ejecta layer have microfloras dominated by one or a few species, usually ferns (Orth et al., 1981; Tschudy et al., 1984; Sweet, 2001; Vajda et al., 2001). This fern-rich assemblage (called the 'fern spike') has been interpreted as an early successional flora that demonstrates the first phases of recovery following the removal of the forest canopy by the blast wave and fire that followed from the bolide impact (Orth et al., 1981; Sweet et al., 1999). Today, ferns are common elements of early successional vegetation, particularly in areas with tropical and subtropical climate, and ferns are also known to be excellent dispersers because their spores can be carried long distances on the wind (Page, 2002).

Although fern spikes are known from multiple K–T boundary sections distributed from New Mexico to Wyoming, fern-dominated assemblages are not found at all K–T boundary sites (Sweet, 2001). Several sections in Canada preserve conifer-dominated microfloras in the section immediately overlying the boundary horizon (Sweet et al., 1999). This seems to demonstrate that recovery from the K–T event had a local component (Sweet, 2001; Nichols and Johnson, 2002). Palynofloral and megafloral data increasingly point to higher survival rates for Cretaceous plant species growing in wetlands (swamps and mires on river floodplains) than for those that grew in better drained habitats (Nichols and Johnson, 2002; Johnson, 2002; Hotton, 2002). Palynofloral data also indicate some taxonomic and ecological selectivity in the K–T extinction in that several groups of probable insect-pollinated plants had a higher rate of extinction than the rest of the flora, as did one group of probable wind-pollinated trees related to oaks (Hotton, 2002). Leaf cuticle fragments from K–T sections in New Mexico and Colorado suggest that broad-leaved evergreen plants related to laurels and avocado had higher rates of extinction than most other types of plants (Wolfe and Upchurch, 1986). Abundance does not appear to have been an important factor in survival because rare and common Cretaceous plants have similar probabilities of extinction at the K–T boundary (Hotton, 2002).

The presence of a fern-dominated flora just above the K–T boundary in New Zealand (Vajda et al., 2001; Figure 3.8) is the first clear evidence that, whatever processes were responsible for causing massive

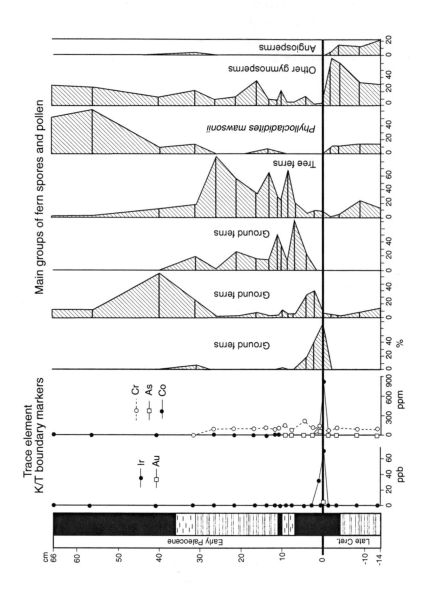

disturbance of terrestrial vegetation at the K–T boundary, they must have operated globally. Global scope is not completely consistent with the impact winter hypothesis because an impact winter would presumably have had a greater effect on the hemisphere experiencing summer conditions at the time of the impact (Wolfe, 1991).

High levels of plant extinction and global scope of devastation are firmly established to have occurred at the K–T boundary, demonstrating that plants are not substantially resistant to the effects of bolide impacts. Less firmly documented, but quite probably true, the extinctions were more severe among broad-leaved evergreen and/or insect-pollinated plants, and possibly among trees, than among small weedy plants. In spite of the probable selectivity of the extinction, there is no evidence that any high-level clade of plants went extinct. Even within flowering plants, which had become by far the most diverse group of land plants well before the end of the Cretaceous (Niklas et al., 1985; Crane and Lidgard, 1989; Lidgard and Crane, 1990; Lupia et al., 1999), there are no documented extinctions of higher taxa. This may reflect, in part, the reluctance of palaeobotanists to name families or orders based on fossils alone, but many extant orders are now known to have originated in the Late Cretaceous (Collinson et al., 1993; Crepet and Nixon, 1998; Magallon et al., 1999; Wing, 2000), and therefore must have survived the bolide impact and subsequent events.

Floral change at the Paleocene–Eocene boundary (c. 55.5 Ma ago)

In the preceding section we have seen that mass extinction of plants was coincident with catastrophic disruption and rapid environmental change at the K–T boundary. The base of the Eocene (55.5 Ma ago) is also marked by a number of geologically rapid events – though of a different and much less destructive type. Over a period of 100 000–200 000 years in the earliest Eocene there was an increase in temperature of 4–8 °C

Figure 3.8. A palynological record from New Zealand showing the dramatic increase in fern spore abundance and decrease in tree pollen (both conifer and angiosperm) following the Cretaceous–Tertiary boundary (modified from Vajda et al., 2001). Note that the conifers recover their proportional abundance slowly, and that the angiosperms remain relatively rare through the entire Paleocene part of the record. Fern dominance of vegetation in this record is estimated to have lasted for 30 000 years.

across mid and high latitudes, a global increase in the amount of the lighter stable isotope of carbon compared to the heavier stable isotope, the biggest ever extinction of bottom-dwelling marine foraminifera, a rapid evolutionary radiation of tropical surface-dwelling foraminifera, and exchange of terrestrial mammal faunas from Asia, North America and Europe across the Bering and North Atlantic land bridges (see articles in Wing *et al.*, 2003a).

The leading hypothesis to explain the occurrence of these events in such a geologically short period is the sudden release of very large quantities of methane that had previously been trapped in ice-like compounds called clathrates contained in ocean-floor sediments (Dickens *et al.*, 1997). Methane is highly enriched in the light isotope of carbon, explaining the worldwide shift in carbon isotope ratios, and is also a powerful greenhouse gas that would help retain heat at the earth's surface. Furthermore, chemical oxidation of methane in the atmosphere would yield two other greenhouse gases, CO_2 and H_2O. The release of methane would also have changed chemical conditions in the ocean, perhaps directly causing the extinction of bottom-dwelling foraminifera. The geologically rapid warming of the Earth's surface (probably in about 10 000 years) is thought to have opened high-latitude land bridges to mammals living in Asia, Europe and North America, thus explaining the intercontinental migration associated with the event (Clyde and Gingerich, 1998). What were the effects of this rapid global warming on plants?

Studies of the floral response to the most recent deglaciation (roughly 20 000–10 000 years ago) have led us to expect that warming of 5–10 °C in mid latitudes will result in rapid, continental-scale shifts in the ranges of many plant species (Jackson and Overpeck, 2000). By analogy with floral change during deglaciation, we might expect to see rapid changes in plant ranges during the earliest Eocene warm period as well. This should show up as higher abundances of tropical taxa, especially at middle and high latitudes. In both surface ocean and terrestrial vertebrate communities, transient faunas and floras are characteristic of the earliest Eocene warm period (Gingerich, 1989; Kelly *et al.*, 1998; Crouch *et al.*, 2001).

The actual record of floral change in the earliest Eocene does not accord with our expectations. Only modest changes in composition are seen in the South American tropics (Rull, 1999; Jaramillo and Dilcher, 2000), North America (Pocknall, 1987; Frederiksen, 1994; Harrington, 2003), and Europe (Jolley, 1998). In these studies, however, the short period of warming in the earliest Eocene was not sampled, so a

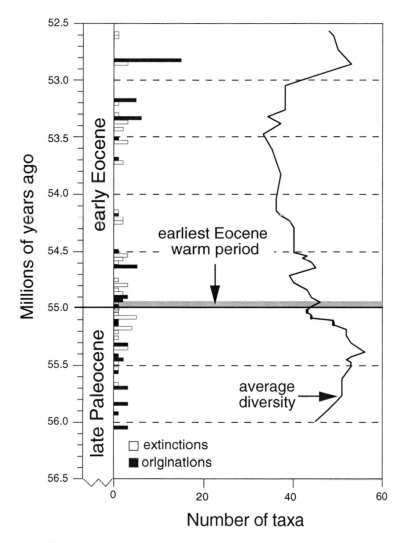

Figure 3.9. Record of plant extinctions and originations across the Paleocene–Eocene boundary interval in Wyoming (modified from Wing, 1998). Although standing diversity declines in the late Paleocene prior to the earliest Eocene warm period, there is no large number of extinctions associated with the warm period. (Recent calibrations of the Cenozoic time scale suggest the age of the Paleocene–Eocene boundary is about 55.5 million years rather than 55.0 million years ago.)

temporary response of floras could have been missed. Recent studies of fossil pollen from New Zealand and North America have sampled the earliest Eocene warm period and show only modest changes in floral composition; there is no indication of a distinctive, transient flora confined to the short warm interval (Crouch, 2001; Crouch and Visscher, 2003; Wing et al., 2003b). Leaf fossils from Wyoming in the USA show that just a few species, mostly ferns, invaded the area during the earliest Eocene warm period, and that few plants went extinct (Wing, 1998; Figure 3.9). Both pollen and leaves show that some types of plants that had previously been abundant became less so, and vice versa. It is interesting that these changes in abundance were not reversed even when temperatures cooled again a few hundred thousand years after the initial warming occurred.

In sum, the events of the earliest Eocene provide an example of major, rapid climatic change that did not cause mass (or even elevated) extinction among plants. The brief warming did not even result in the large-scale changes in geographic range that are associated with rapid climate change during the recent deglaciation of North America and Europe. It is not clear why the pace of floral change was so slow, but the absence of range changes in plants could reflect ecological or physiological limitations (Wing and Harrington, 2001; Wing et al., 2003b).

CONCLUSIONS

This chapter began by identifying three aspects of plant biology that might make their extinction patterns differ from those of animals: plants are individually resistant to physical destruction; they are sessile and therefore vulnerable to climatic change; and they all compete for similar resources – sunlight, water and space to grow in. These aspects of plant biology have been thought to make them unlikely to suffer mass extinctions from bolide impacts, but likely to experience elevated extinction during periods of climatic change or after a new, competitively dominant group evolves. How do these predictions match the case histories reviewed (summarized in Table 3.1)? Below I offer four generalizations.

First, detailed studies of the Permian–Triassic and K–T boundaries suggest that plants experienced extinctions, just as did animals. It is not entirely clear why the extinctions recorded in high-resolution local studies are not observed in broader compilations, such as those of Niklas et al. (1980, 1985), but there are two end-member explanations. One is that the fine-scale studies are recording only local to regional

Table 3.1. A summary of plant extinctions

Event	Cause	Geographic extent	Timing	Severity	Selectivity	Recovery vegetation
Westphalian–Stephanian	Increasing seasonal dryness during deglaciation	Wet tropics of Euramerica	~100 kyr? Diachronous across tropics?	~50–70% megafossil species in swamps, ~40% pollen/spores	Lycopsid swamp trees most affected	Tree ferns (*Psaronius*)
Permian–Triassic	Climate change? Bolide impact?	Global?	~100–200 kyr? Synchronous but plant extinction slightly lags marine?	~97% megafossil species in some Australian sections, 30% pollen/spores	Conifer and glossopterid trees most affected	Small lycopsids (*Pleuromeia*). Slow recovery (~5 Ma)
Cretaceous–Paleogene	Bolide impact, physical destruction, climate change?	Global?	100 years? Synchronous with marine extinction	~80% megafossil species in local sections, ~15–30% pollen/spores	Insect-pollinated and evergreen plants most affected	Varies with locale. Mostly ferns & conifers. Fast recovery (<1 Ma)
Paleocene–Eocene	Climate warming from methane release	Mid to high latitudes?	~150 kyr Synchronous with some marine extinctions	Maximum 10–20% megafossil species in local sections, ~0% pollen/spores	Not applicable	Little extinction but ferns and early successional angiosperms increase after event

Kyr: thousand years; Ma: million years.

events, and that the dramatic changes recorded in these local sections are outweighed in global compilations by data from regions where little happened. This explanation implies that there are not global mass extinctions of plants, just severe local or regional extinctions that happen to coincide with global mass extinctions of animals. An alternative explanation is that long time intervals and inconsistently applied taxonomic names in different areas in the global compilations mask the true declines in diversity. Whatever the global pattern of diversity, it is clear that large proportions of plant species in some regions were eliminated at the Permian–Triassic and K–T boundaries. In the case of the K–T boundary the extinction immediately follows a bolide impact, though it is not yet clear if the fire and blast from the impact are the cause of most of the extinctions, or if the plants were driven to extinction by climatic or other environmental changes that ensued from the impact.

Second, it appears that plant extinctions can be ecologically selective, that is, some types of plants are lost in larger numbers than are others. In all three of the major extinctions discussed, the abundant, large plants appear to have suffered most heavily, lycopsids at the Westphalian–Stephanian boundary, conifers at the Permian–Triassic, and angiosperms at the K–T. Why large plants should have suffered most is not entirely clear, though the reason could be related to the longer life cycles of larger plants. In the aftermath of each extinction there was a period during which small, possibly 'weedy' plants with fast life cycles and good dispersal mechanisms were very important. The durations of these intervals of weed dominance varied enormously, being decades to thousands of years following the K–T extinction, but millions of years following the Permian–Triassic extinction. The spread of weeds in the aftermath of extinctions has been likened to the phenomenon of ecological succession following the disturbance of a forest by fire or storm, but it is clear from the long time involved (even following the K–T boundary) that the processes involved in repopulating landscapes after mass extinctions must be different from those involved in ecological succession.

Third, although large plants are hit harder than smaller plants during extinctions, the clades to which they belong do not go extinct entirely. This is a way in which plant mass extinctions might differ from animal mass extinctions – the ecological selectivity of extinctions may not translate to a high degree of taxonomic selectivity because each major clade of plants shows a considerable diversity of life histories.

Even if the tree members of a clade succumb in a mass extinction, weedier species of the same group may survive and even re-evolve into tree forms in the aftermath. This probably happened during the Permian–Triassic extinction, and perhaps during the K–T extinction as well.

Fourth, not all dramatic and rapid climate changes caused extinctions among plants, as is shown by the events at the Paleocene–Eocene boundary. We are not yet sure why plants experienced so little extinction in connection with the rapid global warming 55 Ma ago, but it may be related to the type of climatic change (warming rather than cooling). Alternatively the magnitude of the change may simply not have been large enough to exceed the tolerances of most species. As with all the other transitions reviewed here, floral changes at the Paleocene–Eocene boundary require further study.

Overall, the response of plants to mass extinctions is different from that of animals, but not as different as we might have thought initially based on their basic biological and ecological differences, and on the apparent absence of mass extinctions in global compilations of plant species through geological time. At the largest scale, the evolutionary histories of animals and plants do look different, because of the survival of most higher taxa of plants. This does not reflect so much the resistance of individual plant species to extinction as it does the development of a very wide range of ways of making a living within each major lineage of land plant. As long as each large group contains some extinction-resistant, weedy lineages, it will survive, with the result that macroevolutionary patterns in plants do not look like those in animals.

SUMMARY

Palaeontologists have identified five global mass extinctions during the Phanerozoic. Each of these episodes sharply reduced the diversity of marine animals, and at the Permian–Triassic, Triassic–Jurassic, and K–T boundaries terrestrial vertebrates also were affected.

These mass extinctions played an important role in shaping the evolutionary history of animals by eliminating some major branches of the tree of life and allowing others to survive. In contrast, plants have been said to be resistant to mass extinction. Global compilations of plant taxonomic data do not reveal any drops of more than about 10 per cent of standing species diversity. Most of the major lineages of vascular plants that evolved during the initial radiation of the group in the Devonian survive to the present day. Even during the two most severe mass extinctions in earth history, at the end of the Permian and the end of the Cretaceous, there is little evidence for extinction of higher taxa of plants.

Although global extinction of higher taxa is rare among plants, detailed studies of fossil plants at some time intervals have revealed evidence of extensive ecological disruption and dramatically decreased plant diversity on both local and regional scales.

Structurally and/or ecologically dominant species were eliminated preferentially during some of these periods and small and/or weedy plants prospered following disruptions. In spite of this, the higher taxa to which the dominant species belonged were not eliminated, and in most cases the formerly dominant lineages eventually gave rise to new dominants.

The role of mass extinctions in the evolutionary history of plants is neither exactly the same as among animals, nor totally different. High resolution records suggest that plants do suffer mass extinctions, just as do animals. The difference is that extinctions among plants largely affect lower taxonomic levels. This implies that extinctions may not to be as taxonomically selective among plants as they are among animals. Lower taxonomic selectivity among plants might result from the broadly similar resource requirement of plant species, or from the high diversity of modes of life within each major clade. A metaphor for the mode of action of mass extinctions in plants as opposed to animals is the effect of a shotgun blast as opposed to a chain-saw. Among plants, mass extinction might act as a shotgun blast, removing many small twigs from the phylogenetic tree, but leaving all major limbs intact. (Even if we imagine that the twigs that are removed share some life history traits, such as having large bodies and slow reproduction.) Regrowth (i.e. diversification) occurs from the same major limbs that were present before the extinction. Mass extinctions among plants would not be followed by evolutionary radiations of higher taxa because the resources released by the extinction are eventually recaptured by species belonging to the same lineages that were present before the extinction took place (sprouts from the same limb). By contrast, among animals mass extinction might act more like a chain-saw that cuts off major limbs of the phylogeny. The 'chain-saw' mode of extinction is easier to detect because all the species of a distinct higher lineage are removed, and because the rediversification of life necessarily comes from different limbs of the tree.

Appendix 3.1   Brief descriptions of the major types of land plants

Although there is evidence that multicellular plants lived on land as early as the Ordovician, and perhaps even the Late Cambrian (Strother, 2000; Edwards and Wellman, 2000), the major evolutionary radiation of vascular plants is thought to have occurred in the Late Silurian–Devonian interval. Relationships among the major lineages of vascular plants are still under study, and remain controversial. The branching diagram, or cladogram, in Figure 3.2 (p. 65) presents one hypothesis of these relationships developed by Kenrick and Crane (1997). Groups in bold are mentioned in the text.

Embryophyta are all living land plants that protect their spore-producing cells, including mosses, liverworts, club mosses, ferns, horsetails and seed plants.

Tracheophyta (vascular plants) have specialized water-conducting cells called tracheids. Tracheids allow plants to grow taller because they move water up the stem more effectively, and in many plants provide mechanical support that keeps the plant from falling over. All living embryophytes with the exception of hornworts, liverworts, and mosses have tracheids and belong to this group. Tracheophytes probably originated in the Silurian.

Lycopsida (club mosses) have a life cycle with free-living sporophyte and gametophyte phases. The sporophyte phase produces sporangia, generally in the axils of small, simple leaves; each sporangium contains numerous small spores that when released can germinate and grow into small, multicellular haploid gametophytes. The gametophytes are small plants that produce either eggs or flagellated sperm that require free water in order to carry out fertilization. The zygote produced by fertilization grows into a diploid sporophyte, thus completing the life cycle. Living lycopsids are small plants – forest floor herbs, epiphytes or rooted aquatics. They typically branch only dichotomously – that is at each branch point two equal branches of smaller diameter are produced. During the Palaeozoic large lycopsid trees were a major component of wetland forests (for example, the scale-tree *Lepidodendron*). The oldest lycopsids are Late Silurian–Early Devonian.

Pteropsida (ferns), like lycopsids, alternate sporophytic and gametophytic generations in their life cycles. Ferns differ from lycopsids in many respects, most obviously by generally having their sporangia on the leaves, and by having large, highly divided leaves.

Some ferns are trees, and these support themselves by growing a mantle of roots from the trunk. Other ferns are forest-floor herbs, floating aquatics, *epiphytes* (plants that grow on other plants), or herbaceous plants of open vegetation. Many ferns are **rhizomatous** – that is they produce many leaves from an underground horizontal stem. Both trees and herbs have evolved convergently in many different lineages of ferns, and both types of architecture have been present since the late Palaeozoic. Ferns make a living in so many different ways that it is difficult to generalize. However, most ferns are good at dispersing because of their small spores, and all are dependent on free water at some point in their life cycles because their sperm are flagellate and must swim to fertilize the egg. The oldest fern-like fossils are Late Devonian.

Sphenopsids (horsetails) also have alternating sporophytic and gametophytic generations, but the sporophytes have a highly distinctive architecture. Upright stems grow from subterranean rhizomes, and bear leaves or branches in whorls. In between the whorls of leaves or branches the stems are filled with soft tissue (**parenchyma**) and air spaces. The rhizome is organized in the same fashion except that it has roots borne in whorls. The sporangia in horsetails and their relatives are borne in terminal clusters that superficially resemble the cones of some seed plants. Living horsetails are commonly found in wet, disturbed settings, and their fossils are common in areas where floods deposited sediment frequently. The relationship of sphenopsids to ferns is disputed, and the oldest sphenopsids are probably also mid-Devonian.

Spermatophytes (seed plants) have megaspores that are surrounded by maternal sporophytic tissue, and the gametophytic phase of the life cycle takes place within this envelope. This is why seed plants tend to be less dependent on water than other vascular plants – there is no phase of the life cycle where free water is required for sperm cells to swim to egg cells. The earliest seed plant fossils are known from the Late Devonian, and are relatively small plants. During the Carboniferous, however, seed plants began to diversify into many life forms, including several types of trees and woody vines. By the Late Permian, virtually all terrestrial vegetation was dominated by seed plants, and they have remained dominant to the present. All of the groups that follow are seed plants.

Medullosales (seed ferns) were a major component of Carboniferous swamp vegetation. They were moderately large trees with very large, fern-like leaves. The trunk was relatively weak and the weight of the crown was borne in part by the thick **petioles**, or leaf stems. The seeds were attached to the leaves, and in some species the seeds were large and surrounded by soft, fleshy tissue that may have been an attractant for animal dispersers, possibly fish or amphibians. The pollen was large and may also have been carried to developing seeds (ovules) by animals. Medullosans went extinct during the Permian.

Cordaitales were an important group of plants in Carboniferous vegetation. Unlike medullosans, cordaites produced dense wood, and bore simple but large strap-shaped leaves. Cordaites ranged from shrubby plants to large, single-stemmed trees. Cordaite seeds were borne in loose, cone-like structures, and indeed cordaites are related to conifers. Most cordaites had winged, presumably wind-dispersed seeds, and some had prop-root systems similar to those seen in living mangrove trees. Like medullosans, cordaites appear to have gone extinct during the Permian.

Cycadales (cycads) have large leaves with multiple leaflets arranged along a central axis. Living cycads are mostly plants with fairly thick trunks composed of soft tissue, and the leaves crowded toward the tip of one or a few stems, but the group had a wider variety of architectures in the past. The (usually large) seeds of cycads are borne either in cones or laterally on leaf-like structures. Many living cycads are pollinated by beetles. The oldest cycad fossils are Early Permian.

Ginkgoales (ginkgos) have a single living species that is commonly planted as a street tree in temperate regions. It has two-lobed, fan-shaped leaves, makes dense wood somewhat like that of conifers, and produces seeds that are terminal on short branches. The seeds are surrounded by a smelly, fleshy covering that may have been involved in attracting dispersers. Fossil ginkgos were variable in the degree of lobing of leaves and in other features. The earliest ginkgo fossils are Permian.

Coniferales (conifers) probably evolved in the Late Carboniferous, but they are rare fossils in lowland deposits until the Permian. From the beginning they probably formed dense wood, bore cones, and had fairly simple, small, scale- or needle-shaped leaves. Most fossil conifers appear to have been trees, as are almost all living species, but an extinct herbaceous conifer *Aethophyllum* is known from the mid-Triassic of France (Rothwell *et al.*, 2000). Many fossil and living conifers appear to be somewhat drought resistant. Conifers diversified extensively during the Permian, and again during the early Mesozoic.

Angiosperms (flowering plants) are the most diverse living group of land plants and were also the last to evolve. The oldest reliable fossil records of the group are fossil pollen grains from the Early Cretaceous. The earliest fossil angiosperms are small plants, possibly rooted aquatics or wetland herbs, but by the mid-Cretaceous the group included various types of herbs, floating aquatic plants, shrubs and trees. Angiosperms also underwent a great diversification of lineages during the mid-Cretaceous, so that by the latter part of the period they were by far the most diverse group of land plants in most regions of the world. In spite of this, conifers and ferns, in particular, remained an important part of vegetation in some areas and habitats (and indeed remain so to the present day).

Appendix 3.2 Megafossils, mesofossils, microfossils and resolving extinctions in the plant fossil record

The fossil record of any extinction is always influenced by where, when and how fossils are preserved. Palaeobotanists recognize three types of plant fossils that tend to produce different kinds of records: **megafossils** (commonly leaves, seeds, and wood), **microfossils** (often referred to as **palynomorphs**, including mostly pollen and spores), and *mesofossils* (small pieces of plants including small flowers and seeds, bits of charcoal, and leaf cuticle). Each type of plant fossil has advantages and disadvantages as a vehicle for studying extinction.

Megafossils are morphologically complex, and because of this it is usually possible to recognize species of plants. Megafossils, however, tend to be preserved in only a restricted set of environments, commonly in wetland, stream or lake deposits. This limits the number of places where studies can be carried out, and usually restricts the **stratigraphic resolution** that can be attained. (If samples are not close together in vertical sections of rock, then it is likely that a long period of time intervened between one sample and the next. Clearly one cannot detect events that take place between samples.) Making large collections of megafossils is also labour-intensive, which tends to limit sample sizes. Furthermore, it is more difficult to assess the composition and diversity of the original vegetation with a small sample. Finally, many megafossil assemblages tend to be highly local, that is, the fossils were not transported and mixed prior to deposition. This means that information about local variation in vegetation is commonly preserved, but also that any individual collection represents only the vegetation immediately around the site. In order to get a good assessment of the plants that were present in a region at a particular time, many localities of the same age have to be collected.

Because they are small, palynomorphs are easily dispersed by the wind over a large area (thousands of square metres to many square kilometres). As a result each pollen sample represents a larger area of vegetation than do typical megafossil samples. Also because they are small, thousands or tens of thousands of palynomorphs can be preserved in a single small piece of rock or sediment, which means that the composition of the ancient flora is better represented in a single sample of palynomorphs. Furthermore, palynomorphs are covered by a decay-resistant coat that allows them to be preserved in a broad range of depositional environments. Together, these features make it possible to sample palynomorphs at high stratigraphic density, and therefore to resolve events that are close together in time. This is an immense advantage when understanding extinctions. There is, however, a big disadvantage to studying palynomorphs: generally one type of pollen or spore represents a whole genus or family of plants. Thus palynomorph records have low **taxonomic resolution** – even if 90 per cent of the species in a genus go extinct, this might not be evident in the palynoflora.

Mesofossils are in many ways intermediate between megafossils and microfossils in how commonly they are preserved, how much area they represent, the degree of taxonomic resolution that can be achieved and in their abundance. As a result they tend to yield intermediate temporal and taxonomic resolution of extinction events.

REFERENCES

Alvarez, L. W., Alvarez, W., Asaro, F. and Michel, H. V., 1980. Extraterrestrial cause for the Cretaceous–Tertiary extinctions. *Science* **208**: 1095–108.
Benton, M. J. (Ed.) 1993. *The Fossil Record*, 2nd edn. London: Chapman and Hall.
Claeys, P., Kiessling, W. and Alvarez, W., 2002. Distribution of Chicxulub ejecta at the Cretaceous–Tertiary boundary. In: C. Koeberl and K. G. MacLeod (Eds.), Catastrophic events and mass extinctions: impacts and beyond. *Geological Society of America Special Paper* **356**: 55–68.
Clark, J. S., Fastie, C., Hurtt, G. *et al.*, 1998. Reid's paradox of rapid plant migration. *BioScience* **48**: 13–24.
Clyde, W. C. and Gingerich, P. D., 1998. Mammalian community response to the latest Paleocene thermal maximum: an isotaphonomic study in the northern Bighorn Basin, Wyoming. *Geology* **26**: 1011–1014.

Collinson, M. E., Boulter, M. C. and Holmes, P. L., 1993. Magnoliophyta ('Angiospermae'). In: M. J. Benton (Ed.), *The Fossil Record*, 2nd edn. London: Chapman & Hall, pp. 809–841.

Crane, P. and Lidgard, S., 1989. Angiosperm diversification and paleolatitudinal gradients in Cretaceous floristic diversity. *Science* **246**: 675–678.

Crepet, W. L. and Nixon, K. C., 1998. Fossil Clusiaceae from the late Cretaceous (Turonian) of New Jersey and implications regarding the history of bee pollination. *American Journal of Botany* **85**: 1122–1133.

Crouch, E. M., 2001. Environmental change at the time of the Paleocene–Eocene biotic turnover. *Laboratory of Palaeobotany and Palynology Contributions Series* **14**: 1–216.

Crouch, E. M., Heilmann-Clausen, C., Brinkhuis, H. *et al.*, 2001. Global dinoflagellate event associated with the late Paleocene thermal maximum. *Geology* **29**: 315–318.

Crouch, E. M. and Visscher, H., 2003. Terrestrial vegetation from across the Initial Eocene Thermal Maximum at the Tawanui marine section, New Zealand. In: S. L. Wing, P. D. Gingerich, B. Schmitz and E. Thomas (Eds.), Causes and consequences of early Paleogene warm climates. *Geological Society of America Special Paper* **369**: 351–364.

Dickens, G. R., Castillo, M. M. and Walker, J. C. G., 1997. A blast of gas in the latest Paleocene: simulating first-order effects of massive dissociation of oceanic methane hydrate. *Geology* **25**: 259–262.

DiMichele, W. A. and Aronson, R. B. 1992. The Pennsylvanian–Permian vegetational transition: a terrestrial analogue to the onshore–offshore hypothesis. *Evolution* **46**: 807–824.

DiMichele, W. A. and Phillips, T. L., 1996. Climate change, plant extinctions, and vegetational recovery during the middle–late Pennsylvanian transition: the case of tropical peat-forming environments in North America. In: M. L. Hart, (Ed.), *Biotic Recovery from Mass Extinctions*. London: Geological Society of London, pp. 201–221.

Edwards, D. and Wellman, C., 2000. Embryophytes on land: the Ordovician to Lochkovian (lower Devonian) record. In: P. G. Gensel and D. Edwards (Eds.), *Plants Invade the Land*. New York: Columbia University Press, pp. 3–28.

Erwin, D. H., Bowring, S. A. and Yugan, J., 2002. End-Permian mass extinctions: a review. In: C. Koeberl and K. G. MacLeod (Eds.), Catastrophic events and mass extinctions: impacts and beyond. *Geological Society of America Special Paper* **356**: 363–383.

Frederiksen, N. O., 1972. The rise of the Mesophytic flora. *Geoscience and Man* **4**: 17–28.

1994. Paleocene floral diversities and turnover events in eastern North America and their relation to diversity models. *Review of Palaeobotany and Palynology* **82**: 225–238.

Gingerich, P. D., 1989. New earliest Wasatchian mammalian fauna from the Eocene of northwestern Wyoming: composition and diversity in a rarely sampled high-floodplain assemblage. *University of Michigan Papers on Paleontology* **28**: 1–97.

Harrington, G. J., 2003. Geographic patterns in the floral response to Paleocene–Eocene warming. In: S. L. Wing, P. D. Gingerich, B. Schmitz, and E. Thomas, E. (Eds.), Causes and consequences of early Paleogene warm climates. *Geological Society of America Special Paper* **369**: 381–394.

Hildebrand, A. R., Penfield, G. T., Kring, D. A. *et al.*, 1991. Chicxulub Crater: a possible Cretaceous/Tertiary boundary impact crater on the Yucatan Peninsula, Mexico. *Geology* **19**: 867–871.

Hotton, C. L., 2002. Palynology of the Cretaceous–Tertiary boundary in central Montana: evidence for extraterrestrial impact as a cause of the terminal Cretaceous extinctions. In: J. H. Hartman, K. R. Johnson and D. J. Nichols (Eds.), The Hell Creek Formation and the Cretaceous–Tertiary boundary in the northern Great Plains: an integrated continental record of the end of the Cretaceous. *Geological Society of America Special Paper* **361**: 473–502.

Jackson, S. T. and Overpeck, J. T., 2000. Responses of plant populations and communities to environmental changes of the Late Quaternary. *Paleobiology* **26**: 194–220.

Jaramillo, C. A. and Dilcher, D. L., 2000. Microfloral diversity patterns of the late Paleocene–Eocene interval in Colombia. *Geology* **28**: 815–818.

Johnson, K. R., 1992. Leaf-fossil evidence for extensive floral extinction at the Cretaceous–Tertiary boundary, North Dakota, USA. *Cretaceous Research* **13**: 91–117.

2002. Megaflora of the Hell Creek and lower Fort Union Formations in the western Dakotas: vegetational response to climate change, the Cretaceous-Tertiary boundary event, and rapid marine transgression. In: J. H. Hartman, K. R. Johnson and D. J. Nichols (Eds.), The Hell Creek Formation and the Cretaceous–Tertiary boundary in the northern Great Plains: an integrated continental record of the end of the Cretaceous. *Geological Society of America Special Paper* **361**: 329–392.

Johnson, K. R. and Hickey, L. J., 1990. Megafloral change across the Cretaceous/Tertiary boundary in the northern Great Plains and Rocky Mountains, U.S.A. In: V. L. Sharpton and P. D. Ward (Eds.), Global catastrophes in earth history: an interdisciplinary conference on impacts, volcanism, and mass mortality. *Geological Society of America Special Paper* **247**: 433–444.

Johnson, K. R., Nichols, D., Attrep, M. J. and Orth, C., 1989. High-resolution leaf-fossil record spanning the Cretaceous–Tertiary boundary. *Nature* **340**: 708–711.

Jolley, D. W., 1998. Palynostratigraphy and depositional history of the Palaeocene Ormesby/Thanet depositional sequence set in southeastern

England and its correlation with continental West Europe. *Review of Palaeobotany and Palynology* **99**: 265–315.

Kelly, D. C., Bralower, T. J. and Zachos, J. C., 1998. Evolutionary consequences of the latest Paleocene thermal maximum for tropical planktonic foraminifera. *Palaeogeography, Palaeoclimatology, Palaeoecology* **141**: 139–161.

Kenrick, P. and Crane, P. R., 1997. *The Origin and Early Diversification of Land Plants: a Cladistic Study.* Washington: Smithsonian Institution Press.

Kerp, H., 2000. The modernization of landscapes during the late Paleozoic–early Mesozoic. In: R. A. Gastaldo and W. A. DiMichele (Eds.), Terrestrial ecosystems, a short course. *The Paleontological Society Papers* **6**: 79–113.

Knoll, A. H., 1984. Patterns of extinction in the fossil record of vascular plants. In: M. H. Nitecki (Ed.), *Extinctions.* Chicago: The University of Chicago Press, pp. 21–68.

Lidgard, S. and Crane, P., 1990. Angiosperm diversification and Cretaceous floristic trends: a comparison of palynofloras and leaf macrofloras. *Paleobiology* **16**: 77–93.

Lomax, B. H., Beerling, D. J., Upchurch, G. R., Jr and Otto-Bliesner, B. L., 2000. Terrestrial ecosystem responses to global environmental change across the Cretaceous–Tertiary boundary. *Geophysical Research Letters* **27**: 2149–2152.

Looy, C. V., Brugman, W. A., Dilcher, D. L. and Visscher, H., 1999. The delayed resurgence of equatorial forests after the Permian–Triassic ecologic crisis. *Proceedings of the National Academy of Science* **96**: 13857–13862.

Lupia, R., Lidgard, S. and Crane, P. R., 1999. Comparing palynological abundance and diversity: implications for biotic replacement during the Cretaceous angiosperm radiation. *Paleobiology* **25**: 305–340.

Magallon, S., Crane, P. R. and Herendeen, P. S., 1999. Phylogenetic pattern, diversity, and diversification of eudicots. *Annals of the Missouri Botanical Garden* **86**: 297–372.

Nichols, D. J. and Fleming, R. F., 1990. Plant microfossil record of the terminal Cretaceous event in the western United States and Canada. In: V. L. Sharpton and P. D. Ward (Eds.), Global catastrophes in earth history: an interdisciplinary conference on impacts, volcanism, and mass mortality. *Geological Society of America Special Paper* **247**: 445–456.

Nichols, D. J. and Johnson, K. R., 2002. Palynology and microstratigraphy of Cretaceous–Tertiary boundary sections in southwestern North Dakota. In: J. H. Hartman, K. R. Johnson and D. J. Nichols (Eds.), The Hell Creek Formation and the Cretaceous–Tertiary boundary in the northern Great Plains: an integrated continental record of the end of the Cretaceous. *Geological Society of America Special Paper* **361**: 95–144.

Niklas, K. J., 1997. *The Evolutionary Biology of Plants.* University of Chicago Press: Chicago and London.

Niklas, K. J. and Tiffney, B. H., 1994. The quantification of plant biodiversity through time. *Philosophical Transactions of the Royal Society, London, Series B* **345**: 35–44.

Niklas, K. J., Tiffney, B. H. and Knoll, A. H., 1980. Apparent changes in the diversity of fossil plants. *Evolutionary Biology* **12**: 1–89.

1985. Patterns in vascular land plant diversification: an analysis at the species level. In: J. W. Valentine (Ed.), *Phanerozoic Diversity Patterns: Profiles in Macroevolution*. Princeton: Princeton University Press, pp. 97–128.

Norris, R. D. and Firth, J. V., 2002. Mass wasting of Atlantic continental margins following the Chicxulub impact event. In: C. Koeberl and K. G. MacLeod (Eds.), Catastrophic events and mass extinctions: impacts and beyond. *Geological Society of America Special Paper* **356**: 79–95.

O'Keefe, J. D. and Ahrens, T. J., 1989. Impact production of CO_2 by the Cretaceous–Tertiary extinction bolide and the resultant heating of the Earth. *Nature* **338**: 247–249.

Orth, C., Gilmore, J., Knight, J., Pillmore, C., Tschudy, R. and Fassett, J., 1981. An iridium abundance anomaly at the palynological Cretaceous–Tertiary boundary in northern New Mexico. *Science* **214**: 1341–1342.

Page, C. N., 2002. Ecological strategies in fern evolution: a neopteridological overview. *Review of Palaeobotany and Palynology* **119**: 1–33.

Pocknall, D. T., 1987. Paleoenvironments and age of the Wasatch Formation (Eocene), Powder River Basin, Wyoming. *Palaios* **2**: 368–376.

Pollack, J. B., Toon, O. B., Ackerman, T. P., McKay, C. P. and Turco, R. P., 1983. Environmental effects of an impact-generated dust cloud: implications for the Cretaceous–Tertiary extinctions. *Science* **219**: 287–289.

Pope, K. O., 2002. Impact dust not the cause of the Cretaceous–Tertiary mass extinction. *Geology* **30**: 99–102.

Pope, K. O., Baines, K. H., Ocampo, A. C. and Ivanov, B. A., 1994. Impact winter and the Cretaceous/Tertiary extinctions: results of a Chicxulub asteroid impact model. *Earth and Planetary Science Letters* **128**: 719–725.

Rees, P. M., 2002. Land-plant diversity and the end-Permian mass extinction. *Geology* **30**: 827–830.

Retallack, G. J., 1995. Permian–Triassic life crisis on land. *Science* **267**: 77–80.

Retallack, G. J., Veevers, J. J. and Morante, R., 1996. Global coal gap between Permian–Triassic extinction and Middle Triassic recovery of peat-forming plants. *Geological Society of America Bulletin* **108**: 195–207.

Rothwell, G. W., Grauvogel-Stamm, L. and Mapes, G., 2000. An herbaceous fossil conifer: gymnospermous ruderals in the evolution of Mesozoic vegetation. *Palaeogeography, Palaeoclimatology, Palaeoecology* **156**: 139–145.

Rowley, D., Raymond, A., Parrish, J., Lottes, A., Scotese, C. and Ziegler, A., 1985. Carboniferous paleogeographic, phytogeographic, and paleoclimatic reconstructions. *International Journal of Coal Geology* **5**: 7–42.

Rull, V., 1999. Palaeofloristic and palaeovegetational changes across the Paleocene/Eocene boundary in northern South America. *Review of Palaeobotany and Palynology* **107**: 83–95.

Schultz, P. H. and D'Hondt, S., 1996. Cretaceous–Tertiary (Chicxulub) impact angle and its consequences. *Geology* **24**: 963–967.

Sepkoski, J. J., Jr, 1993. Ten years in the library: new data confirm paleontological patterns. *Paleobiology* **19**: 43–51.

Shen-Miller, J., Mudgett, M. B., Schopf, J. W., Clarke, S. and Berger, R. 1995. Exceptional seed longevity and robust growth: ancient sacred lotus from China. *American Journal of Botany* **82**: 1367–1380.

Strother, P. K., 2000. Cryptospores: the origin and early evolution of the terrestrial flora. In: R. A. Gastaldo and W. A. DiMichele (Eds.), Phanerozoic terrestrial ecosystems. *The Paleontological Society Special Papers* **6**: 3–20.

Sweet, A. R., 2001. Plants, a yardstick for measuring the environmental consequences of the Cretaceous–Tertiary boundary event. *Geoscience Canada* **28**: 127–138.

Sweet, A. R., Braman, D. R. and Lerbekmo, J. F., 1999. Sequential palynological changes across the composite Cretaceous–Tertiary (K–T) boundary claystone and contiguous strata, western Canada and Montana, USA. *Canadian Journal of Earth Sciences* **36**: 743–768.

Traverse, A., 1988. Plant evolution dances to a different beat: plant and animal evolutionary mechanisms compared. *Historical Biology* **1**: 277–301.

Tschudy, R., Pillmore, C., Orth, C., Gilmore, J. and Knight, J., 1984. Disruption of the terrestrial plant ecosystem at the Cretaceous–Tertiary boundary, Western Interior. *Science* **225**: 1030–1032.

Twitchett, R. J., Looy, C. V., Morante, R., Visscher, H. and Wignall, P. B., 2001. Rapid and synchronous collapse of marine and terrestrial ecosystems during the end-Permian biotic crisis. *Geology* **29**: 351–354.

Vajda, V., Raine, J. I. and Hollis, C. J., 2001. Indication of global deforestation at the Cretaceous–Tertiary boundary by New Zealand fern spike. *Science* **294**: 1700–1702.

Valentine, J. W. 1985. *Phanerozoic Diversity Patterns: Profiles in Macroevolution.* Princeton: Princeton University Press.

Visscher, H., Brinkhuis, H., Dilcher, D. L. *et al.*, 1996. The terminal Paleozoic fungal event: evidence of terrestrial ecosystem destabilization and collapse. *Proceedings of the National Academy of Sciences* **93**: 2155–2158.

Wing, S. L., 1998. Late Paleocene–early Eocene floral and climatic change in the Bighorn Basin, Wyoming. In: W. Berggren, M. P. Aubry and S. Lucas (Eds.), *Late Paleocene–Early Eocene Biotic and Climatic Events.* New York: Columbia University Press, pp. 380–400.

   2000. Evolution and expansion of flowering plants. In: R. A. Gastaldo and W. A. DiMichele (Eds.), Terrestrial ecosystems, a short course. *The Paleontological Society Papers* **6**: 209–232.

Wing, S. L. and Harrington, G. J., 2001. Floral response to rapid warming in the earliest Eocene and implications for concurrent faunal change. *Paleobiology* **27**: 539–562.

Wing, S. L., Gingerich, P. D., Schmitz, B. and Thomas, E. (Eds.) 2003a. Causes and consequences of early Paleogene warm climates. *Geological Society of America Special Paper* **369**, 624 pp.

Wing, S. L., Harrington, G. J., Bowen, G. J. and Koch, P. L., 2003b. Floral change during the Initial Eocene Thermal Maximum in the Powder River Basin, Wyoming. In: S. L. Wing, P. D. Gingerich, B. Schmitz and E. Thomas (Eds.), Causes and consequences of early Paleogene warm climates. *Geological Society of America Special Paper* **369**: 425–440.

Wolbach, W. S., Gilmour, I., Anders, E., Orth, C. J. and Brooks, R. R., 1988. Global fire at the Cretaceous–Tertiary boundary. *Nature* **334**: 665–669.

Wolfe, J. A., 1990. Palaeobotanical evidence for a marked temperature increase following the Cretaceous–Tertiary boundary. *Nature* **343**: 153–156.

  1991. Palaeobotanical evidence for a June 'impact winter' at the Cretaceous/Tertiary boundary. *Nature* **352**: 420–423.

Wolfe, J. A. and Upchurch, G. R., Jr, 1986. Vegetation, climatic and floral changes at the Cretaceous–Tertiary boundary. *Nature* **324**: 148–152.

DAVID J. BOTTJER

Department of Earth Sciences, University of Southern California,
Los Angeles, USA

4

The beginning of the Mesozoic: 70 million years of environmental stress and extinction

INTRODUCTION

Coral reefs are one of Earth's environments where a visitor experiences sensory overload. A snorkeler or diver in a tropical shallow sea is commonly overwhelmed with a dazzling array of colourful fish and invertebrate animals, including corals of all shapes and sizes (Figure 4.1). Corals are colonial animals and the many polyps of the colony build large common skeletons made of calcium carbonate that typically constitute the main physical structure of modern reefs. Lurking in the nooks and crannies of coral reefs are a great variety of animals with fascinating morphologies and life habits (Figure 4.1). It is no wonder that coral reefs are such a popular destination for human recreation.

The fascination of coral reefs is not only for the casual observer, but also for scientists interested in the numbers of animals on Earth and general issues of biodiversity. Coral reefs are found along one-sixth of the world's coastlines (Birkeland, 1997), and are the most biologically diverse of nearshore ecosystems (Roberts *et al.*, 2002). They are an integral part of the natural landscape on Earth today, and yet they are one of the ecosystems that has been most severely affected by global warming and human activities. Thus, much like tropical rain forests, coral reefs are considered to be fragile environments that we are in danger of losing due largely to our own activities. It is hard for anyone to imagine a world without coral reefs.

And, yet, surprisingly, there have been times in Earth's history when reefs made by corals and other colonial animals largely disappeared from the surface of the Earth for periods of several million years. These were typically times of stress for Earth's ocean environments, and

Extinctions in the History of Life, ed. Paul D. Taylor.
Published by Cambridge University Press. © Cambridge University Press 2004.

Figure 4.1. A large variety of organisms characterize modern coral reefs. Components of diversity include colonial organisms such as corals and sponges as well as fish and boring organisms that live inside coral skeletons (inset). Modified from Stanley (1992, 2001) with permission.

indicate that as in modern environments such reefs have also acted in the past as 'canaries in the mine' for the ocean's ecosystems. We will examine the long-term history of reefs as we start our journey into the beginning of the Mesozoic.

REEFS DURING THE BEGINNING OF THE MESOZOIC

The time in Earth's history that is marked by proliferation of animal life is termed the Phanerozoic Eon, and this is divided into the Palaeozoic (543–251 million years (Ma) ago), the Mesozoic (251–65 Ma ago), and the Cenozoic (65 Ma ago to today). The Mesozoic, which includes three time periods – the Triassic (251–200 Ma ago), the Jurassic (200–135 Ma ago), and the Cretaceous (135–65 Ma ago) – was preceded by the greatest mass extinction suffered by Earth's animals, and ended with another great mass extinction, which led to the extinction of the last species of dinosaurs.

To the trained eye, fossil reefs are fairly easily identified in the stratigraphic record. Reefs with a significant component of colonial organisms first evolved at the beginning of the Palaeozoic (Cambrian, 543–495 Ma ago), when an unusual group of sponges (archaeocyathans) were the primary component, and in the following Ordovician time period bryozoans and corals joined sponges to form types of reefs that endured through much of the rest of the Palaeozoic (Wood, 1999). Intriguingly, during the first 70 million years of the Mesozoic, from the Early Triassic through the Early Jurassic, there are two long intervals of geological time when reefs made by colonial animals were rare to non-existent, the Early Triassic and the Early Jurassic (Stanley, 1988, 2001) (Figure 4.2).

The Early Triassic 'reef eclipse' is a striking feature in the history of life. Reefs in the Permian (295–251 Ma ago), the last period of the Palaeozoic, were large and were characterized by a variety of colonial animals, primarily sponges and bryozoans. These reef-building organisms were victims of the end-Permian mass extinction, where over half of the families of marine organisms became extinct (Chapter 5). While Early Triassic reefs made by colonial animals are absent, in Lower Triassic strata of south China, the Middle East, and the western United States, small patch reefs made of calcium carbonate by microbial organisms are found (Lehrman, 1999; Pruss and Bottjer, 2001; Baud et al., 2002). The Early Triassic represents a period of biotic recovery from the end-Permian mass extinction (Bottjer, 2001), and formation only of small microbial reefs may reflect high levels of environmental stress for marine animals. Estimates for duration of this time period range from 4 to 10 Ma (Martin et al., 2001).

It was only well into the Middle Triassic, perhaps as long as 12 Ma after the end-Permian mass extinction (Stanley, 2001), that tropical reef ecosystems characterized by colonial organisms began to reappear (Figure 4.2). These first Mesozoic reefs with a large contribution from colonial organisms included sponges and bryozoans. Scleractinian corals, which are the major contributor to most of today's reefs, first appeared during the Middle Triassic, but did not assume a major role in reefs until some time in the Late Triassic, perhaps as much as 30 Ma after the end-Permian mass extinction (Stanley, 2001).

Reefs built by colonial animals such as corals and sponges during the Late Triassic (230–200 Ma ago) represent one of the great reef-building episodes in Earth history (Figure 4.2). This all came to an end with another of Earth's great mass extinctions, that at the end of the Triassic (200 Ma ago) (Chapter 5). During the first few million years of the Jurassic, no reefs made by colonial animals are known (Figure 4.2). A few are found in the middle of the Early Jurassic (200–180 Ma ago), but they then disappear at the end of the Early Jurassic, only to reappear in the Middle Jurassic (180–154 Ma ago). This approximately 20 Ma-long

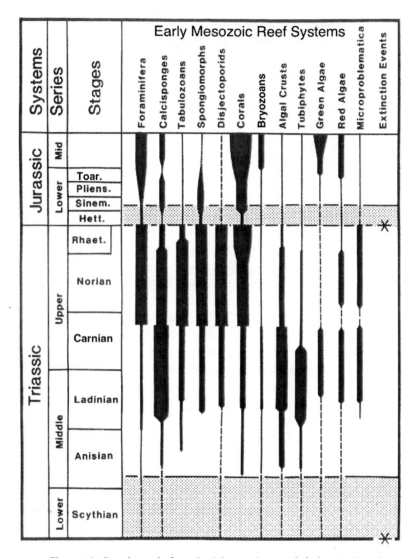

Figure 4.2. Broad trends for colonial organisms and their associates in early Mesozoic reef systems. Widths of bars indicate relative importance of individual reef organisms. Series and stages are subdivisions of Triassic and Jurassic time (Rhaet., Rhaetian; Hett., Hettangian; Sinem., Sinemurian; Pliens., Pliensbachian; Toar., Toarcian). Asterisks indicate the end-Permian and end-Triassic mass extinctions, with stippled intervals showing times of reef eclipse. Modified from Stanley (1988) with permission.

period in the beginning of the Jurassic is marked by a minor mass extinction (early Toarcian; 183 Ma) before the second disappearance of reefs made by colonial animals. Interestingly, as for the Early Triassic, it has recently been discovered that other organisms moved into the role of reef-builders during the Early Jurassic 'reef eclipse'. These Early Jurassic reefs are constructed mainly by aberrant large bivalve molluscs (*Lithiotis*), which are found only during this Early Jurassic time of biotic recovery from mass extinctions (Fraser and Bottjer, 2001). The occurrence of such reefs made by bivalves is also thought to indicate that there existed a high level of environmental stress for marine organisms. The re-establishment of reefs built primarily by colonial metazoans in the Middle Jurassic include those dominated by corals and sponges (Stanley, 2001) (Figure 4.2).

It is likely that, as at the present day, the loss of reefs made by colonial animals from ancient environments indicates stressed ocean ecosystems for those intervals of Earth history. Thus, this outline of early Mesozoic reef history shows that for much of the beginning of the Mesozoic, ecosystems in the world's oceans were experiencing significant environmental stress.

OTHER BIOLOGICAL INDICATORS OF EARLY
MESOZOIC CONDITIONS

The biological participants in deep waters of open oceans lack the glamour of tropical reef ecosystems. Much of the biological activity in such environments is by free-floating plankton, especially tiny single-celled organisms known as protists. The fossil record of such protists at the end of the Palaeozoic and the beginning of the Mesozoic is dominated by radiolarians, which build their skeletons of silica (glass). Radiolarians have a long and rich fossil record that begins in the Cambrian. Some seafloor-inhabiting sponges also build skeletons of silica. When sediments are formed into sedimentary rocks through such processes as compaction, these silica skeletons are commonly dissolved and re-precipitated as the rock type 'chert'. Intriguingly, studies of Lower Triassic sedimentary rocks show a dramatic drop in the diversity of radiolarians and siliceous sponges, and the lack of chert (Racki, 1999). The cause of this 'chert gap' is poorly understood, although it was undoubtedly related to the environmental changes that drove the end-Permian mass extinction.

Although we have been concentrating on evidence indicating the condition of Early Triassic oceans, there is much to learn from the

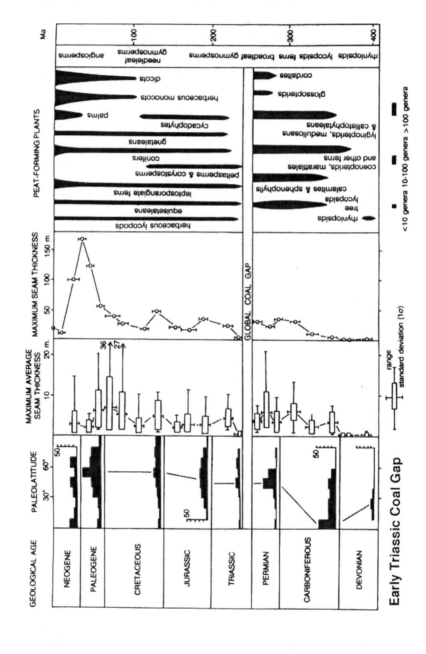

Early Triassic Coal Gap

land. Wetland ecosystems that form peat have existed on Earth since the Devonian (410–355 Ma ago) (Retallack et al., 1996). When peat is transformed by compaction and associated processes into a sedimentary rock it becomes coal. Coals are found in sedimentary rocks from the Devonian onward, except during the Early Triassic (Retallack et al., 1996) (Figure 4.3). This Early Triassic 'coal gap' is followed by the occurrence of only a few thin and uncommon coals in the Middle Triassic (Figure 4.3). Thus, for a time period as long as 20 Ma after the end-Permian mass extinction wetland ecosystems that form peat were absent to extremely rare (Retallack et al., 1996; Chapter 3).

Early Jurassic biotic change is not as grand as that seen in the Early Triassic. But, still, the end-Triassic mass extinction not only affected reefs but organisms of the open ocean, with the extinction of the conodonts, tiny chordates with a long and major fossil record, as well as the near extinction of the ammonites, which were swimming shelled cephalopod molluscs and one of the most abundant macrofossil groups of the Mesozoic (Hallam and Wignall, 1997).

A variety of other palaeoecological evidence shows that the Early Triassic and Early Jurassic were times of ecological degradation (e.g. Ward, 2000; Bottjer, 2001). Data on taxonomic diversity powered the first line of studies on extinctions (e.g. Raup and Sepkoski, 1982; Sepkoski, 1993; Benton, 1995), and continued analysis of such data has shown that extinction during the Triassic is not just concentrated at the beginning and the end of this time, but that all of the Triassic is typified by relatively high extinction rates (Bambach and Knoll, 2001) (Figure 4.4).

CAUSES OF LONG-TERM ECOLOGICAL DEGRADATION

What could have caused this long-term ecological degradation during the early Mesozoic? To find some possible clues on today's Earth for the cause we need to travel to Iceland (Figure 4.5). There we will go to Laki, where evidence can be found for the only flood basalt fissure

←—————————————————————————————————

Figure 4.3. Broad trends in aspects of coal deposition since the Devonian. Palaeolatitudes of coals indicate latitudinal changes in coal deposition for particular geological time intervals. Similarly, coal seam thickness indicates thickness of coal sedimentary beds including maximum values of unusually thick seams for particular time intervals (to right) and averages for at least 10 coal seams within a sedimentary sequence (to left). Dominant plants are shown in column next to ages (Ma = million years). Modified from Retallack et al. (1996) with permission.

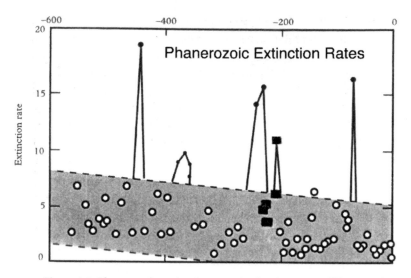

Figure 4.4. Phanerozoic extinction rates (extinctions per million years) for families of marine invertebrates and vertebrates. Each data point is for a stratigraphic stage, each of which represents a relatively short interval of geological time (see Figure 4.2 for examples). Filled circles and square above shaded region represent major mass extinctions (from left to right) during end-Ordovician, Late Devonian, end-Permian, end-Triassic and end-Cretaceous times (all except Late Devonian are statistically significant). Filled squares represent extinction rates calculated for Triassic stages. Shaded area shows trend of declining background extinction during the Phanerozoic. Identification of minor mass extinction intervals is based on additional considerations (Hallam and Wignall, 1997). Note that for the 251 million years after the end-Permian mass extinction four of the eight highest extinction rates are recorded for intervals in the Triassic. Modified from Raup and Sepkoski (1982) and Courtillot (1999) with permission.

eruption that has occurred in historical times. We go there because climatic and associated environmental effects of flood basalt volcanism in the past have been assigned causal roles for several mass extinctions, particularly those marking the beginning of the Mesozoic (e.g. Hallam and Wignall, 1997; Courtillot, 1999; Wignall, 2001; Chapter 5).

On June 8 of 1783 a large rift opened up near Laki and soon lava flowed from the rift at a rate of 7200–8700 m^3 per second (Gudmundsson, 1996). In July, as eruptions issued from new fissures, lava fountains from 800–1400 m high formed. The eruption continued along the ultimately 25-km-long fissure until May, 1785, resulting in a volume of 14–15 km^3, including 0.75 km^3 of tephra, which today covers

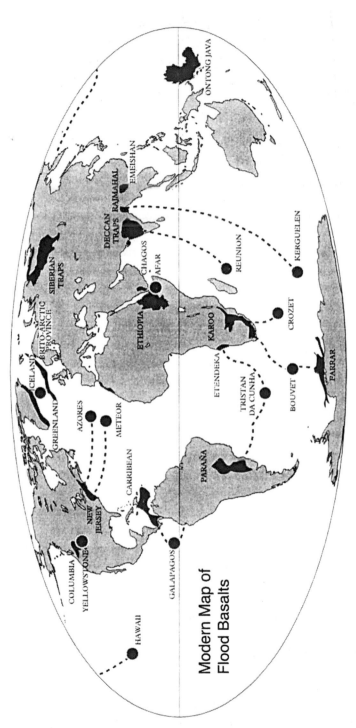

Figure 4.5. Modern map of the main flood basalts (large igneous provinces) on Earth deposited through geological time. Areal distribution of flood basalts shown in black. Dashed lines link flood basalts to currently active mantle plumes (circles) which may have caused those flood basalts in the distant past, when the continent on which the flood basalt is currently found was located over the mantle plume. Mantle plumes remain relatively stationary, while the continents are transported by plate tectonic processes. Iceland is currently forming due to eruptions from a mantle plume. Modified from Courtillot (1999) with permission.

an area of 580–600 km². The climatic impact included the release of *c.* 140 million tonnes of sulphur dioxide that formed at least 80 million tonnes of sulphuric acid, which led to a bluish haze in the atmosphere. Crop failures in Iceland during the summer of 1783, as well as the exceptionally warm summer of 1783 in northern Europe and cold winter for 1783–1784 in the northern hemisphere, are all thought to have been caused by the Laki eruptions (e.g. Gudmundsson, 1996; Wignall, 2001). These geographically widespread effects are of great interest because the Laki eruption was a relatively small one when compared with those found in the geological record.

Indeed, in the past, tremendous outpourings of flood basalt, much greater than the individual event recorded at Laki, have created on Earth what have been termed 'large igneous provinces' (LIPS) (Figure 4.5). One of the largest of these is the Siberian Traps, which occurs today in an area of western Siberia totalling 3.9 million km², and which originally could have included as much as 2.6 million km³ of basalt (Reichow *et al.*, 2002). The Siberian Traps formed through eruptions that occurred over several million years straddling the Permian–Triassic boundary of 251 Ma ago (Wignall, 2001; Reichow *et al.*, 2002) (Figure 4.6). Detrimental environmental effects of the eruption of the Siberian Traps, as well as subsequent stressful climatic and oceanographic conditions likely to have been caused by such voluminous outpourings of flood basalt and associated volcanics, have been considered by many to be strong candidates for the cause of the end-Permian mass extinction (e.g. Renne and Basu, 1991; Reichow *et al.*, 2002).

At the end of the Permian, the Earth looked much different than it does today. By 300 Ma all of the continents had assembled into one landmass, Pangea, which extended from pole to pole (Figure 4.6). If it had been possible to view Earth from outer space then, one side would have been all Pangea, and the other side would have been completely covered by ocean. The assembly of Pangea had been completed through collisions driven by plate tectonics which had sutured together many continental plates over several hundred million years. Huge outpourings of flood basalts are generally thought to be driven by mantle plumes, which ascend from deep within the mantle to erupt on the surface, and are well-known for creating chains of islands in the ocean such as Hawaii (Figure 4.5).

The Siberian Traps may have been generated by such a mantle plume (Wignall, 2001). Plumes underneath continents, a type of LIP termed a 'continental flood basalt province' (CFBP), can initiate continental rifting, which ultimately leads to the continent breaking up

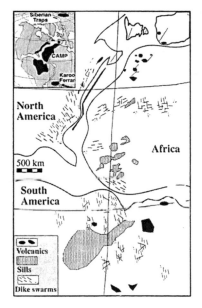

Early Mesozoic Pangean Continental Flood Basalt Provinces

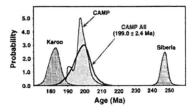

Figure 4.6. Evidence for distribution and age of early Mesozoic Pangean continental flood basalt provinces. Geological evidence (occurrence of volcanics, sills and dike swarms) for central Atlantic magmatic province (CAMP), as well as geography of Pangea and distribution of Siberian Traps, Karoo Traps, and Ferrar Traps (inset) are shown on left. On right are age probability spectra for radiometrically-determined age dates of samples from CAMP, the Siberian Traps and the Karoo Traps. The curve labelled CAMP includes only those samples from CAMP with minimal determined analytical error. Modified from Marzoli et al. (1999) with permission. Ma, million years ago.

with an ocean forming between the two remaining pieces. It is possible that the Siberian Trap CFBP was formed by a mantle plume that failed to break up Pangea. However, recent work shows the presence of an enormous CFBP along the margins of today's Atlantic Ocean in eastern North America, northwest Africa, and northeast South America (Figure 4.6). This Central Atlantic Magmatic Province (CAMP) covers an area of 7 million km^2 and may have also had an original volume of 2 million km^3 (Marzoli et al., 1999). CAMP volcanism was probably caused by a mantle plume that initiated a breakup of Pangea which eventually led to the formation of the central Atlantic Ocean (e.g. Marzoli et al., 1999). The volcanism occurred over a period of several million years, with a peak probably at 200 Ma ago, which corresponds to the age of the end-Triassic mass extinction (Marzoli et al., 1999)

(Figure 4.6). As for the end-Permian mass extinction, climatic effects of CAMP volcanism, as well as related stressful climatic and oceanographic conditions caused by this volcanism, have been suggested as the cause of the end-Triassic mass extinction (e.g. Wignall, 2001; Hesselbo et al., 2002).

The Karoo Traps in South Africa and the Ferrar Traps in Antarctica indicate emplacement of another CFBP that led to further breakup of Pangea and formation of the South Atlantic (e.g. Wignall, 2001) (Figures 4.5, 4.6). These two traps also contain an enormous volume of volcanics, approximately 2.5 million km^3, and show a concentration of eruptions in the early Toarcian, an interval of the Early Jurassic about 183 Ma ago (Wignall, 2001). This is the time when another mass extinction occurred, although it was of less significance than those at the end of the Permian and the end of the Triassic. As for these other two extinctions, this one has also been postulated to have been caused by this South Atlantic volcanism (Hesselbo et al., 2000).

What might be the long-term effects of such large episodes of volcanic eruptions? As described for Laki, we know a little from historical records about the short-term effects of flood basalt volcanism on the weather. But there is really nothing in human experience to understand the long-term effects on climate of many eruption episodes as big or much bigger than Laki, such as occurred to form each of the three Pangean CFBPs during the beginning of the Mesozoic. However, we do know that CO_2 is an important greenhouse gas, that it is injected into the atmosphere as part of volcanic eruptions, and that such CO_2 lingers in surficial systems for a relatively long time (e.g. Wignall, 2001) (Figure 4.7). Thus, CO_2 generated through formation of CFBPs during the beginning of the Mesozoic could have substantially warmed the climate during this time.

Work is ongoing to try to understand atmospheric CO_2 levels, as well as global temperatures, for the beginning of the Mesozoic, and for other times even deeper in Earth history (e.g. McElwain et al., 1999; Wignall, 2001; Berner and Kothavala, 2001; Crowley and Berner, 2001; Tanner et al., 2001; Beerling, 2002; Retallack, 2002) (Figure 4.8). Depending upon the approach used results are not always consistent, but a variety of studies show that the Triassic through the Early Jurassic is a time of both increased atmospheric CO_2 (Figure 4.8) and global temperatures (Crowley and Berner, 2001). Thus, it seems likely that the Earth was significantly warmer during this time, and that continental flood basalt volcanism could have been the cause of this long-term warming.

Increased warming of the Earth's atmosphere will lead to a decreased temperature gradient from the equator to the poles. This

Duration of Volcanic Effects

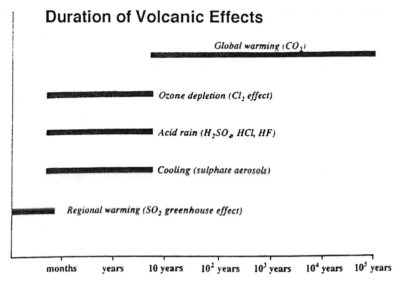

Figure 4.7. Duration of the effects of volcanic gasses. Other than CO_2 most volcanic gasses are removed from the atmosphere in a matter of years and thus affect only weather, whereas the global warming from increased CO_2 can affect long-term climate. Modified from (Wignall 2001) with permission.

probably would lead to decreased overall circulation of the Earth's oceans, which in turn would cause stagnation and reduced oxygen content of ocean waters below surface layers. The Earth had a giant ocean – Panthalassa – during this time, which surrounded the supercontinent Pangea. Studies of sedimentary rocks that were deposited on the deep-sea floor of the Panthalassa Ocean, now found outcropping in Japan, show the existence of a long period of time when deep ocean waters had little to no dissolved oxygen (Isozaki, 1997). These conditions, termed 'anoxia', indicate great stagnation of the deep ocean from the end of the Permian into the beginning of the Middle Triassic, a period of time 20 Ma long (Isozaki, 1997). Evidence for anoxic conditions in shallower ocean waters adjacent to the continents also exists, but this anoxia lasted for a relatively shorter time, from the end of the Permian into the early part of the Early Triassic (Hallam and Wignall, 1997). Anoxia in the oceans is also associated with the early Toarcian extinction event of the Early Jurassic (Hallam and Wignall, 1997; Hesselbo et al., 2000).

The combination of global warming and oceanic anoxia would increase the stress for organisms living on land and in the sea. Thus, evidence from the geological record demonstrates that the beginning of the Mesozoic was a time of long-term environmental stress (e.g. Woods

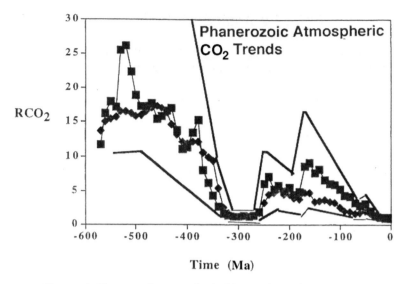

Time (Ma)

Figure 4.8. Phanerozoic atmospheric CO_2 trends, as determined using the GEOCARB II and III models. These models are based on inputs of geological, geochemical, biological and climatological data. RCO_2 is the ratio of CO_2 levels with respect to the present (300 parts per million), time is in millions of years (Ma). Results from GEOCARB II are plotted with diamonds, results from GEOCARB III are plotted with squares, outer lines represent an estimate of errors in the GEOCARB III model. Results for both models show a similar pattern: very high CO_2 values in the early Phanerozoic, a drop beginning c. 450 Ma ago that bottomed out at c. 300 Ma ago, then an increase beginning at c. 250 Ma ago that continued through the early Mesozoic, followed by a gradual decrease beginning at 170 Ma ago until present times. Modified from Berner and Kothavala (2001) with permission.

et al., 1999; Hesselbo et al., 2000), with the least stressful time of this 70-Ma-long interval being the c. 25-Ma-long interval of the Late Triassic. The significant environmental stress for much of the Triassic may have caused prolonged stress of the biota and ultimately may be why there are relatively high overall extinction rates for the Triassic (Bambach and Knoll, 2001) (Figure 4.4).

CAUSES OF EARLY MESOZOIC MASS EXTINCTIONS

Superimposed on this long period of environmental stress are two major (end-Permian, end-Triassic) (Figure 4.4) and one minor (early Toarcian) mass extinctions (Hallam and Wignall, 1997). Mass extinctions require relatively intense periods of environmental stress over a geologically short interval of time (1 day to 100 000 years) to produce the elevated extinction levels that are used to identify them. A popular cause postulated for all three of these mass extinctions is a relatively intense period of volcanic activity during formation of the CFBPs that were developing during the time of each of these mass extinctions (e.g. Renne and Basu, 1991; Hallam and Wignall, 1997; Courtillot, 1999; Marzoli et al., 1999; Hesselbo et al., 2000, 2002; Isozaki, 2001, Wignall, 2001) (Figure 4.6). Such intense volcanic periods would produce long-term global warming and oceanic anoxia as well as a host of associated effects, including short-term cooling (if eruption of volcanic dust and sulphate aerosols was great enough), acid rain, and release of gas hydrates from the seafloor (e.g. Wignall, 2001) (Figure 4.7).

The other equally effective mechanism that could cause widespread environmental stress during a very short time interval is extraterrestrial bolide impact. This is the mechanism that is strongly implicated for the mass extinction at the end of the Mesozoic (end-Cretaceous) (Figure 4.4), which caused the extinction of the last dinosaurs (e.g. Alvarez et al., 1980; Hallam and Wignall, 1997). Evidence in support of a bolide impact at the end of the Mesozoic includes concentrations of iridium, shocked quartz, and microtektites at the boundary and the presence of a large crater of the right age (c. 65 Ma ago) in the Yucatan Peninsula of Mexico (Hallam and Wignall, 1997). Despite an intensive search of sedimentary successions that include the Palaeozoic–Mesozoic boundary, no evidence like that which occurs at the boundary ending the Mesozoic has been found, so that it has generally been concluded that bolide impact was not involved in the end-Permian mass extinction. However, several recent studies have offered new types of geochemical data that could support the hypothesis that the end-Permian mass extinction was caused by an asteroid or cometary impact (Becker et al., 2001; Kaiho et al., 2001; but see also Koeberl et al., 2002). Similarly, for the end-Triassic mass extinction, there has been a report indicating the presence of shocked quartz at the Triassic–Jurassic boundary (Bice et al., 1992), but present support for an extraterrestrial cause is not strong. No evidence has been found to suggest that a bolide impact led to the extinctions of the early Toarcian.

IMPLICATIONS

Much of the work to date on the effects of extinction during the early Mesozoic has concentrated on the brief time intervals where there are mass extinctions. In large part these mass extinctions have been treated exclusively with little attempt to investigate whether the broader events of this time helped shape these mass extinctions and their impact on life.

The synthesis that has been presented here provides a pathway upon which to re-direct the focus of research on extinction affecting the early Mesozoic. This broad outline indicates that the beginning of the Mesozoic was a time of unusually long-term environmental stress, likely to be due to the beginning of the breakup of Pangea, from intensive mantle plume volcanism that formed several huge CFBPs (Figure 4.6). Superimposed upon these long-term stressful intervals were short-term punctuations of environmental stress that caused three mass extinctions which affected the first 70 Ma of the Mesozoic. Causes of these mass extinctions might have been coupled with the instigator of long-term stress, beginning with unusually intense periods of flood basalt volcanism, that could have led to a potential cascade of unusually stressful tectonic, oceanographic and climatic effects. The cause could also have been decoupled from formation of CFBPs, which for this case would most likely be an extraterrestrial asteroid or cometary impact.

Whatever the ultimate cause of these three mass extinctions that mark the early Mesozoic, the pattern presented here touches upon some intriguing questions about the ways that bolide impacts may affect the biota here on Earth. Some large impacts that have been documented from the geological record have apparently had little effect upon Earth's biota. For example, during the Late Triassic, the major time of the early Mesozoic when a CFBP was not forming, there was a large bolide impact in Quebec, Canada that produced the Manicouagan crater, which has been dated as about 214 Ma old (Hallam and Wignall, 1997), and which has a diameter of 90 km (Becker, 2002). However, no mass extinction is reported from this time. Environmentally, the Late Triassic was the most benign time of the early Mesozoic (e.g. see Figures 4.2, 4.8), and it is possible that the Manicouagan impact did not have severe consequences for the biota because it occurred when Earth's ecosystems were robust enough to adsorb the blow. Perhaps, for bolide impacts that have a severe and measurable impact of extinction on Earth's biota, an Earth that is already under significant environmental stress could be tipped into a mass extinction through the added stress caused by the impact. Thus, if the Earth's environments are already stressed, this may 'set the table' for a mass extinction if a bolide impact were to occur.

These sorts of linkages have begun to intrigue scientists studying mass extinctions not only at the beginning of the Mesozoic, but also during the Palaeozoic and the end of the Mesozoic (Racki and Wrzolek, 2001). For example, the end-Cretaceous mass extinction, which occurred

65 Ma ago at the end of the Mesozoic, along with strong evidence for a large extraterrestrial impact, also occurred during the formation of a CFBP, the Deccan Traps in India (Racki and Wrzolek, 2001) (Chapter 5). For a long time, scientists pursued the approach that either extraterrestrial impact or the effects of the Deccan Trap volcanism caused this mass extinction. Clearly an impact occurred, and the broad consensus is that this impact was the primary cause of the mass extinction. However, as for extinctions affecting the early Mesozoic, several workers on the end-Cretaceous mass extinction have begun to propose that perhaps the Deccan Trap volcanism and the associated environmental stress 'set the table' for the impact that caused the Chicxulub Crater in the Yucatan to tip the Earth's ecosystems into an interval of severe mass extinction (Racki and Wrzolek, 2001).

Such ideas take us back to our modern Earth, where, as has already been discussed, we know ecosystems are under unusual stress due to global warming and other human-induced environmental stress. Ever since the discovery that an extraterrestrial impact caused the extinction of the dinosaurs, we have been fascinated with the possibility that a large asteroid or comet impact could occur with devastating consequences. We are not experiencing the environmental stress that extensive eruption of a CFBP could bring, but it is likely that the level of stress which human activities have placed on natural ecosystems has put them perilously close to mass extinction if an impact, or some Earth-bound sudden oceanographic or climatic change, were to occur.

CONCLUSIONS

The Mesozoic began about 251 Ma ago after the greatest mass extinction that animal life has ever experienced on Earth. The effects of this extinction lingered for millions of years into the Triassic, the first time period of the Mesozoic. Life returned to 'normal' later in the Triassic, only to suffer new crises and extinction in a variety of environments through the early part of the succeeding Jurassic period. This 70 Ma interval of heightened environmental stress and biotic crisis may have been due to the same underlying cause. The root mechanism is very likely that this was the time of the early breakup of the supercontinent Pangea and formation of three huge CFBPs, which led to stressful environmental conditions causing the long interval of biotic crises and extinction that characterizes the beginning of the Mesozoic. If extraterrestrial impacts were the cause of one or more of the mass extinctions that marks the early Mesozoic, it is likely that the already stressed environments of much of this time 'set the table' for the Earth's biota to enter a phase of mass extinction. The beginning of the Mesozoic represents a crucial time in the evolution of life on Earth, because survivors of this prolonged period of environmental stress founded much of the animal life that characterizes today's oceans.

ACKNOWLEDGEMENTS

Bill Schopf is acknowledged for support and encouragement through my long association with CSEOL. Ideas espoused in this paper have been greatly improved through discussion and interaction with Nicole Bonuso, Frank Corsetti, Mary Droser, Alfred Fischer, Margaret Fraiser, Nicole Fraser, Karina Hankins, Tran Huynh, Catherine Jamet, Pedro Marenco, George McGhee, Sara Pruss, David Rodland, Jennifer Schubert, Peter Sheehan, Carol Tang, Richard Twitchett and Adam Woods.

REFERENCES

Alvarez, L. W., Alvarez, W., Asaro, F. and Michel, H. V., 1980. Extraterrestrial cause for Cretaceous–Tertiary extinction. *Science* **208**: 1095–1108.

Bambach, R. K. and Knoll, A. H., 2001. Is there a separate class of 'mass' extinctions? *Geological Society of America Abstracts with Programs* **33**: A-141.

Baud, A., Richoz, S., Cirilli, S. and Jarcoux, J., 2002. Basal Triassic carbonate of the Tethys: a microbialite world. In: Knopen, M. and Cairncross, B. (Eds.), *16th International Sedimentological Congress, Abstract Volume*. Johannesburg: Rand Afrikaans University, pp. 24–25.

Becker, L., 2002. Repeated blows. *Scientific American* **286**: 76–83.

Becker, L., Poreda, R. J., Hunt, A. G., Bunch, T. E. and Rampino, M., 2001. Impact event at the Permian–Triassic boundary: evidence from extraterrestrial noble gases in fullerenes. *Science* **291**: 1530–1533.

Beerling, D., 2002. CO_2 and the end-Triassic mass extinction. *Nature* **415**: 386–387.

Benton, M. J., 1995. Diversification and extinction in the history of life. *Science* **268**: 52–58.

Berner, R. A. and Kothavala, Z., 2001. Geocarb III: a revised model of atmospheric CO_2 over Phanerozoic time. *American Journal of Science* **301**: 182–204.

Bice, D. M., Newton, C. R., McCauley, S. E., Reiners, P. W. and McRoberts, C. A., 1992. Shocked quartz at the Triassic–Jurassic boundary in Italy. *Science* **255**: 443–446.

Birkeland, C. (Ed.), 1997. *Life and Death of Coral Reefs*. New York: Chapman and Hall.

Bottjer, D. J., 2001. Biotic recovery from mass extinctions. In: Briggs, D. E. G. and Crowther, P. R. (Eds.), *Palaeobiology II*. Oxford: Blackwell Science, pp. 202–206.

Courtillot, V., 1999. *Evolutionary Catastrophes*. Cambridge: Cambridge University Press.

Crowley, T. J. and R. A. Berner., 2001. CO_2 and climate change. *Science* **292**: 870–872.

Fraser, N. M. and Bottjer, D. J., 2001. The beginning of Mesozoic bivalve reefs: Early Jurassic 'Lithiotis' facies bioherms. *PaleoBios* **21** (Suppl. to No. 2): 54.

Gudmundsson, A., 1996. *Volcanoes in Iceland*. Reykjavik: Vaka-Helfafell.

Hallam, A. and Wignall, P. B., 1997. *Mass Extinctions and their Aftermath*. Oxford: Oxford University Press.

Hesselbo, S. P., Grocke, D. R., Jenkyns, H. C., Bjerrun, C. Y., Faermond, P., Morgans Bell, H. S. and Green, O. R., 2000. Massive dissociation of gas hydrate during a Jurassic oceanic anoxic event. *Nature* **406**: 392–395.

Hesselbo, S. P., Robinson, S. A., Surlyk, F. and Piasecki, S., 2002. Terrestrial and marine extinction at the Triassic–Jurassic boundary synchronized with major carbon-cycle perturbation: a link to initiation of massive volcanism? *Geology* **30**: 251–254.

Isozaki, Y., 1997. Permo-Triassic boundary superanoxia and stratified superocean: records from lost deep sea. *Science* **276**: 235–238.

     2001. Plume winter scenario for the Permo-Triassic boundary mass extinction. *PaleoBios* **21** (Suppl. to No. 2): 570–71.

Kaiho, K., Kajiwara, Y., Nakano, T. *et al.*, 2001. End-Permian catastrophe by bolide impact: evidence of a gigantic release of sulfur from the mantle. *Geology* **29**: 815–818.

Koeberl, C., Gilmour, I., Reimold, W. U., Claeys, P. and Ivanov, B., 2002. End-Permian catastrophe by bolide impact: evidence of a gigantic release of sulfur from the mantle: Comment. *Geology* **30**: 855–856.

Lehrman, D. J., 1999. Early Triassic calcimicrobial mounds and biostromes of the Nanpanjiang Basin, south China. *Geology* **27**: 359–362.

Martin, M. W, Lehrman, D. J., Bowring, S. A. *et al.*, 2001. Timing of Lower Triassic carbonate bank buildup and biotic recovery following the end-Permian extinction across the Nanpanjiang Basin, south China. *Geological Society of America Abstracts with Programs* **33**: A201.

Marzoli, A., Renne, P. R., Piccirillo, E. M., Ernesto, M., Bellieni, G. and De Min, A., 1999. Extensive 200 million-year-old continental flood basalts of the Central Atlantic Magmatic Province. *Science* **284**: 616–618.

McElwain, J. C., Beerling, D. J. and Woodward, F. I., 1999. Fossil plants and global warming at the Triassic–Jurassic boundary. *Science* **285**:1386–1389.

Pruss, S. B. and Bottjer, D. J., 2001. Development of microbial fabrics in Early Triassic oceans. *PaleoBios* **21** (Suppl. to No. 2): 106.

Racki, G., 1999. Silica-secreting biota and mass extinctions: survival patterns and processes. *Palaeogeography, Palaeoclimatology, Palaeoecology* **154**: 107–132.

Racki, G. and Wrzolek, T., 2001. Causes of mass extinctions. *Lethaia* **34**: 200–202.

Raup, D. M. and Sepkoski, J. J., Jr, 1982. Mass extinctions in the fossil record. *Science* **215**: 1501–1503.

Reichow, M. K., Saunders, A. D., White, R. V. *et al.*, 2002. 40Ar/39Ar dates from the West Siberian Basin: Siberian flood basalt province doubled. *Science* **296**: 1846–1849.

Renne, P. R. and Basu, A. R., 1991. Rapid eruption of the Siberian Traps flood basalts at the Permo-Triassic boundary. *Science* **253**: 176–179.

Retallack, G. J., 2002. Triassic–Jurassic atmospheric CO_2 spike. *Nature* **415**: 387–388.

Retallack, G. J., Veevers, J. J. and Morante, R., 1996. Global coal gap between Permian–Triassic extinction and Middle Triassic recovery of peat-forming plants. *Geological Society of America Bulletin* **108**: 195–207.

Roberts, C. M., McClean, C. J., Veron, J. E. N. *et al.*, 2002. Marine biodiversity hotspots and conservation priorities for tropical reefs. *Science* **295**: 1280–1284.

Sepkoski, J. J., Jr, 1993. Ten years in the library: new data confirm paleontological patterns. *Paleobiology* **19**: 43–51.

Stanley, G. D., Jr, 1988. The history of early Mesozoic reef communities: a three step process. *Palaios* **3**: 170–183.

1992. Tropical reef ecosystems and their evolution. In: Nierenberg, W. A. (Ed.), *Encyclopedia of Earth System Science*. New York: Academic Press, pp. 375–388.

2001. Introduction to reef ecosystems and their evolution. In: Stanley, G. D., Jr (Ed.), *The History and Sedimentology of Ancient Reef Systems*. New York: Kluwer Academic/Plenum Publishers, pp. 1–39.

Tanner, L. H., Hubert, J. F., Coffey, B. P. and McInerney, D. P., 2001. Stability of atmospheric CO_2 levels across the Triassic/Jurassic boundary. *Nature* **411**: 675–677.

Ward, P. D., 2000. *Rivers in Time*. New York: Columbia University Press.

Wignall, P. B., 2001. Large igneous provinces and mass extinctions. *Earth-Science Reviews* **53**: 1–33.

Wood, R., 1999. *Reef Evolution*. Oxford: Oxford University Press.

Woods, A. D., Bottjer, D. J. Mutti, M. and Morrison, J., 1999. Lower Triassic large sea floor carbonate cements: their origin and a mechanism for the prolonged biotic recovery from the end-Permian mass extinction. *Geology* **27**: 645–648.

PAUL B. WIGNALL

School of Earth Sciences, University of Leeds, UK

5

Causes of mass extinctions

WHAT ARE MASS EXTINCTIONS?

It has long been appreciated that the rates of extinction recorded in the fossil record have not remained constant through time, but only during the last few decades has this variation been more clearly quantified. Much of this has been due to the single-handed efforts of the late Jack Sepkoski of the University of Chicago. Sepkoski spent 'ten years in the library' sifting through palaeontological publications and amassing a vast database on the first and last appearances of fossil groups (Sepkoski, 1994). Initially this work concentrated on families of organisms, but it was subsequently up-dated to include the first and last appearances of genera. Plotting last appearances (extinctions) against time revealed several distinctive features (Figure 5.1). Firstly extinction rates appear to have been considerably higher in the earlier part of the fossil record, particularly in the Cambrian Period. This is, at least partly, an artefact of the way extinction rates are measured. Diversity in the Cambrian was relatively low, particularly compared with the levels achieved in the Mesozoic and Cenozoic, with the result that relatively few organisms needed to go extinct to achieve a relatively high extinction percentage (see also Chapter 1).

The other clear signal to emerge from Sepkoski's compilations is that there have been five intervals when extinction rates have greatly exceeded background rates of extinction, these are the 'Big Five' mass extinctions of the fossil record (Figure 5.1). The best known of these is of course the extinction at the end of the Cretaceous, generally known as the K-T event, when the dinosaurs disappeared along with many other

Extinctions in the History of Life, ed. Paul D. Taylor.
Published by Cambridge University Press. © Cambridge University Press 2004.

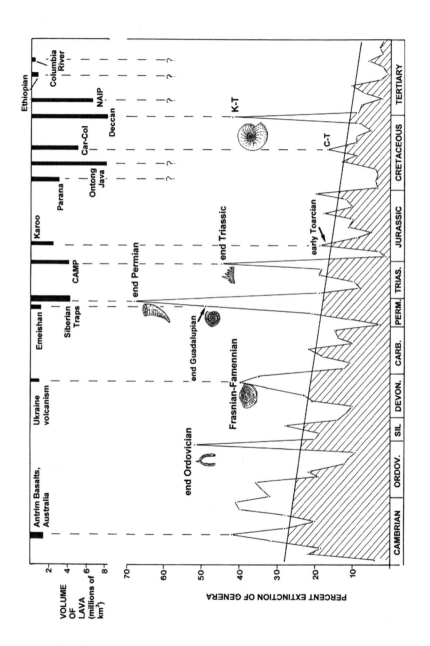

groups including the ammonites. However, in terms of sheer magnitude the biggest event of the fossil record occurred at the end of the Permian, about 250 million years (Ma) ago, when in excess of 90 per cent of all species of animals and plants became extinct. The other three mass extinctions, which occurred at the end of the Ordovician, in the Late Devonian (known as the Frasnian–Famennian (F–F) event because it occurs at the boundary between these two Late Devonian stages) and at the end of the Triassic, were not of this magnitude but nonetheless record severe crises in the history of life.

Several lesser extinction events are also known from the fossil record and some of these have also been designated mass extinctions, notably one in Early Jurassic (the early Toarcian event) and one in the Late Cretaceous (the Cenomanian–Turonian boundary event). However, the only one of these 'lesser' extinction events of comparable magnitude to the 'Big Five' occurred at the end of the Middle Permian. This was only recognized during the 1990s, primarily from a study of fossiliferous marine rocks in China. The global nature of this event is unclear due to the difficulty of precisely age-dating rocks of Permian age and correlating them around the world, but its magnitude is such that the Big Five should now be termed the 'Big Six'. This is the end-Middle Permian mass extinction and is variously known as the end-Guadalupian event (named after the stage name generally used in North America) or the end-Maokouan event (named after the stage name generally used in China and much of Asia) (Stanley and Yang, 1994).

Mass extinctions record geologically brief intervals (less than a million years) of global crisis that affected both terrestrial and marine environments. Understanding their cause has become a topic of paramount importance to the palaeontological community and far beyond. Thus, many current geopolitical and environmental concerns can be found in these studies, and in some instances have been directly triggered by this work, a case of politicians being influenced by palaeontologists! Most notable is the doomsday scenario of nuclear winter

←————————————————————————————————————

Figure 5.1. Plot of generic-level per cent extinction during the Phanerozoic, with main extinctions events named (data from Sepkoski, 1994). A fitted linear least-square regression line illustrates the marked decline of extinction intensity through time. Extinction peaks are compared with the volume of basalt associated with large volcanic provinces plotted long the same time scale revealing a very close correspondence. ORDOV., Ordovician; SIL, Silurian; DEVON., Devonian; CARB., Carboniferous; PERM., Permian; TRIAS., Triassic.

following full-scale nuclear war, which derives much of its credibility from the postulated effects of the K–T meteorite impact. Concerns about the burning of fossil fuels and the emission of greenhouse gases would perhaps not have so much influence were it not for the observation that past episodes of rapid global warming often correspond with mass extinctions. The attempts to unravel the causes behind the catastrophes form the theme of this chapter.

THE NATURE OF THE EVIDENCE

Proving what caused an event that happened many millions of years ago is far from easy. Cause and effect can only be postulated, with the best proof coming from the relative timings. Obviously, the cause should not happen after the effect, but this relies on high precision dating (ideally to within a few tens of thousands of years) which is not always achievable. A frequent correlation of events and effects is also strongly suggestive that a proposed extinction hypothesis is along the right lines. This is one of the main lines of argument used for the volcanic kill mechanism for example, but of course correlation need not necessarily indicate causation. The correlation principle also ignores the fact that some mass extinctions may have unique causes. Thus, giant meteorite impact has only convincingly been demonstrated to be the cause of one mass extinction event (see below).

Demonstrating the cause(s) of mass extinctions generally requires two things: finding a proximate cause and an ultimate kill mechanism. This requires two theories to be formulated, with the ultimate kill mechanism interpreted to be a consequence of the proximate cause. This is clearly demonstrated by the seminal work of Alvarez *et al.* (1980), which kick-started the study of mass extinctions from a relatively minor concern in the palaeontological community to one at the centre of the field. Alvarez and colleagues proposed that the ultimate kill mechanism of the K–T mass extinction was a giant meteorite impact (or bolide impact – the two terms are synonymous), and they also proposed that the proximate cause of the extinction was a global dust cloud that caused the shutdown of photosynthetic activity. Obviously, the latter may have been a consequence of the former but it is important to remember that these are two separate ideas, albeit closely linked. Less linkage is seen in other mass extinction scenarios. For example, marine anoxia is strongly implicated as the proximate cause of the end-Permian extinction in the seas, but the link with an ultimate cause, often thought to be volcanic activity, is only weakly established.

The study of ultimate causes of extinctions is generally concerned with the discovery of a 'smoking gun' – clear, categorical evidence that an event happened. The giant impact crater at Chicxulub (Yucatan, Mexico) is a very big 'smoking gun' for the K–T extinction, as are the vast provinces of flood basalts that coincide with many extinction events. Other proximate causes, such as supernova explosions, require more indirect, some would say more tenuous, evidence that is often of a chemical nature. More circumstantial evidence comes from the fossil record. Thus, the abrupt nature of the K–T extinction suggests a geologically instantaneous catastrophe – a prediction of a meteorite impact cause – but this is not proof of the impact.

META EORITE IMPACT

End-Cretaceous impact

The gradual gathering of evidence for a giant meteorite impact at the end of the Cretaceous forms one of the most fascinating geological detective stories of recent times (Alvarez, 1997). The recognition of a giant impact crater at Chicxulub in the Yucatan Peninsula in the late 1980s provides near-irrefutable evidence of the event, but the idea of an impact predates this discovery by nearly a decade. An enrichment of iridium in K–T boundary clays from Gubbio in Italy and Stevns Klint in Denmark (Figure 5.2) provided the first clue. Iridium, like other platinum group metals, is very rare in the Earth's crust because it sank into the core during the Earth's formation. It is, however, more abundant in many types of meteorite. Therefore, when iridium concentrations reach several parts per billion in a sedimentary rock it is often called an 'anomaly', and it is taken to indicate an input of meteoritic material, but the tricky point is determining if this material was delivered to the Earth in one go, by a very big meteorite, or over a long period of time by the numerous tiny meteorites that bombard our planet all the time as a kind of cosmic rain. Alvarez and colleagues (1980) convincingly argued that the iridium anomaly occurred in sediments that had not accumulated unusually slowly and so were unlikely to record the background rain of micrometeoritic material; instead, sudden introduction from a very big meteorite was implied. By calculating the amount of iridium in the boundary clay, and with knowledge of typical iridium concentrations in meteorites, they suggested a 10-km-diameter meteorite struck the Earth at the end-Cretaceous. Cosmic objects move at between 10 and 80 km/s so the violence of the impact would have been

Figure 5.2. Cretaceous–Tertiary boundary section at Stevns Klint, near Copenhagen, Denmark. The actual boundary occurs at the base of a thin dark clay that interrupts the pale limestones of this cliff section. Clay samples from this location provided some of the first high iridium values that gave evidence for a giant meteorite impact.

of unimaginable magnitude. Measurement of the size of the Chicxulub crater at around 180 km (Hildebrandt *et al.*, 1995), provides a further independent measure of the size of the meteorite, which suggests that the original estimate of Alvarez *et al.* (1980) was remarkably accurate (see Carlisle, 1995).

The K–T iridium anomaly is strong evidence for meteorite impact, although early critics of the idea suggested that a volcanic origin could not be ruled out. However, increasing attention to K–T boundary sediments revealed a plethora of further lines of evidence. Small grains of shocked quartz were discovered soon after 1980 and provided a valuable additional clue (Bohor *et al.*, 1984). Such grains show multiple cross-cutting dark bands, most unlike typical quartz grains; these are lamellae formed by deformation under the extremely high pressures characteristic of meteorite impact events. Around the same time, boundary sediments from around the world were found to contain spherules, tiny spheres up to 100 μm in size, composed of either feldspar or iron and nickel-rich spinel; the latter is an unusual discovery because spinels are rare in most meteorites, although they may be a more common comet constituent (Carlisle, 1995). Through the 1980s more and more different types of evidence for impact were discovered in thin sediment layers at the K–T boundary, including tiny diamonds with cosmic carbon isotope ratios, tektites (glassy melt rock thrown out from the impact), stishovite (high pressure variant of quartz) and various lines of geochemical evidence. The discovery of the Chicxulub crater was really just the icing on the cake that finally silenced many doubters.

The other main aspect of the Alvarez *et al.* (1980) paper, the kill mechanism, has also withstood the test of time and investigation remarkably well. They argued that all the dust thrown up from the impact site would cause global darkness for many years. The consequent shut-down of photosynthesis on land and in the oceans would ensure the starvation of entire ecosystems. In the marine realm, such conditions of zero primary productivity have been termed 'Strangelove Oceans' (Hsu and McKenzie, 1985), and both carbon isotope and palaeontological evidence suggests that they happened briefly at the K–T boundary (see section below), although the case is better established in equatorial rather than at higher latitudes (Hallam and Wignall, 1997).

A further consequence of the Chicxulub impact may have come from the radiative heating of the Earth's surface as impact material re-entered the atmosphere, having been blown out of the crater. Calculations suggest a brief pulse of heating of perhaps several hundred degrees. In fact this aspect of the extinction scenario is so lethal that one wonders how any terrestrial life survived.

Other impacts

Of the remaining five mass extinctions, meteorite impact has been suggested as a cause of three others (the Frasnian–Famennian (F–F), end Permian, and end Triassic events). Examining each in turn reveals that the evidence in each case is, at best, highly tentative. During the Late Devonian several meteorites appear to have hit the Earth, with the largest impact occurring at Siljan in Sweden where a 52 km-diameter crater is found. This crater, together with two other somewhat smaller craters, is approximately 367 Ma old. Unfortunately for proponents of impact-induced mass extinction this is somewhere between 3 and 10 Ma too early to be implicated in the great F–F mass extinction. No craters or impact evidence are known from the F–F boundary. Undaunted, several workers have proposed that there may have been a very long lag period between the multiple impact events and the climate changes that caused the extinction (McGhee, 2002). Judge for yourself whether you consider that the Earth's climate has a 'memory' lasting millions of years, but remember that the effects of the Chicxulub impact were instantaneous!

No impact crater of end-Permian age has been found, and it would need to be a very large crater indeed to be implicated in the greatest mass extinction of all time. Instead the evidence is based on a few rather suspect grains of shocked quartz from an Antarctic locality (Retallack et al., 1998) and a few tiny grains composed of iron–nickel–silica from a single Chinese location, thought to be derived from a meteorite (Kaiho et al., 2001). Recently, trace amounts of fullerenes have been found in sediments from two Permian–Triassic boundary locations (Becker et al., 2001). Fullerenes are unusual organic molecules composed of spherical cages of 60 carbon atoms; they are thought to be present in comets, although they can also be produced by terrestrial processes such as lightning strikes. The end-Permian examples occur in sediments that purportedly yield light helium atoms (^3He) typical of extraterrestrial sources. However, independent analysis of the samples of Becker and her colleagues has failed to find the helium, and one of the two studied locations turns out not to be at the Permian–Triassic boundary at all. On such flimsy evidence is the impact–extinction model built. Physical evidence for meteorite impact coincident with the end-Triassic mass extinction is equally sparse: shocked quartz grains from an Italian section in which the level of the extinction and the position of the Triassic–Jurassic boundary are not known (Hallam and Wignall, 1997). A 100-km-diameter crater at Manicouagan in Canada was tentatively implicated in this extinction, but dating evidence indicates it is at least 10 Ma too old. This raises the question of what happened 10 Ma before the end of the Triassic? The answer is nothing. This interval was in the Norian Stage, a time of notably low extinction rates.

The relationship between meteorite impacts and extinctions is thus rather an intriguing one. One of the largest extinction events of the fossil record was probably caused by an impact but other appreciable-sized impacts, such as the one at Manicougan and the several examples in the Late Devonian, had no effect. What was so lethal about the Chicxulub impact? The fact that it was the biggest impact event in the last 600 Ma years may be important; some sort of lethal threshold may have been crossed whereby the devastation was global and prolonged. Secondly, the target rocks at Chicxulub were limestones and evaporites (precipitates of the sea such as the salt calcium sulphate), which would have been vaporized to release vast volumes of CO_2 and SO_2 into the atmosphere with potentially catastrophic climatic effects (see volcanic section below). The Manicougan impact in comparison occurred in ancient rocks of the Canadian Shield which would merely have sent a lot of rock dust into the atmosphere. The dinosaurs appear to have been killed-off by a very large meteorite hitting the wrong place: bad luck indeed!

Other extraterrestrial causes

The meteorite impact model invokes an extraterrestrial cause for the ultimate kill mechanism, but it is a cause that leaves ample physical evidence that can be discovered, as we have seen. However, from time to time other extraterrestrial causes are invoked as kill mechanisms, for example passage of dark matter through the Earth or bombardment by cosmic rays from a nearby supernova. The difficulty with such ideas is proving them. Supernovae at least generate a range of exotic heavy elements (e.g. ^{244}Pu and ^{129}I) but their concentrations in terrestrial sediments are never likely to be high and their analysis is exceedingly difficult. Carlisle (1995) provides a useful discussion of the search for such elements at the K–T boundary, but the investigations so far have drawn a blank.

MASSIVE VOLCANISM

Within a couple of years of the publication of the seminal Alvarez *et al.* (1980) paper, a rival kill mechanism for the end-Cretaceous extinction was proposed: Deccan Traps volcanism. The Traps occur in northwest India and are just one manifestation of a voluminous style of volcanism that produces extensive sheets of lava known as flood basalts. Something of the scale of this type of volcanism may be appreciated by considering the size of flows compared to historical eruptions and their monitored effects. The largest historical eruption occurred at Laki in Iceland between 1783 and 1784 when 14 km^3 of lava were erupted from a fissure (see also Chapter 4). The environmental effects were devastating for Iceland, and detectable throughout much of the northern hemisphere. Sulphur dioxide emissions appear to have done the most damage. This gas reacts with water vapour to form clouds of sulphuric acid that block out the sun's light and, after a year or so, is rained out of the atmosphere as acid rain. Thus, Benjamin Franklin noted that the summer of 1784 was unusually cold (he was, at the time, on diplomatic service in Paris), a fact that he attributed, with remarkable accuracy, to volcanic activity. The scale of the Laki eruption is trivial compared with individual flows in continental flood basalt provinces, which range from several hundred to as much as 10 000 km^3 (Courtillot, 1999). Flood basalt provinces contain several hundred such flows with the result that their total volume often exceeds a million (10^6) km^3. The Deccan Traps are an especially large province with an original volume perhaps as high as 4 × 10^6 km^3.

The key factor that implicated the Deccan Traps in the end-Cretaceous extinction was improvements in dating of the province. This revealed two things. Firstly the majority of the flood basalt flows were shown to have been erupted in a very short interval of time, perhaps as little as a million years, thus ensuring that the time for recovery between the giant eruption events would have only been a few thousand years. Secondly, this brief interval was shown to straddle the K–T boundary. The Deccan Traps are just one of several giant volcanic provinces that have formed in the past 300 million years (Figure 5.1). Older examples have probably occurred but the lavas have long since been eroded away. This raises the possibility that, if one flood basalt province caused a mass extinction, then perhaps the others also did. This was just what was proposed by Rampino and Stothers in 1988 when they produced an infamous diagram (or famous, depending on your viewpoint) in which they plotted the ages of extinction events against the ages of the flood basalt provinces (Figure 5.3). Plotting two time series against each other inevitably produces a positive correlation, and so is not a very good way of proving anything. Rampino and Stothers also had to apply a considerable 'fudge factor' by inventing several mass extinction events that do not exist (for example in the Palaeocene and the Oligocene). Neither was the age of many volcanic provinces known to the accuracy depicted in Figure 5.3. Thus, it is perhaps not surprising that the viewpoint that all mass extinctions coincide with massive volcanism did not gain general acceptance. Raup (1991, p. 153) succinctly summarized the prevailing view a few years after Rampino and Stothers's proposal:

> There are approximately six really large deposits of flood volcanism in the geologic record, including the Columbia River basalts and the Deccan. There is a hint of an age correspondence with mass extinction in several cases. For example, Deccan volcanism may have started at almost the time of the Cretaceous extinction. Unfortunately, other examples are so few and dating uncertainties so great that good statistical testing is impossible. Thus, the proponents of flood volcanism (and there a good number of them) have had much difficulty convincing their colleagues.

Over a decade later the situation has changed remarkably and revealed Rampino and Stother's idea to have much going for it. The change has

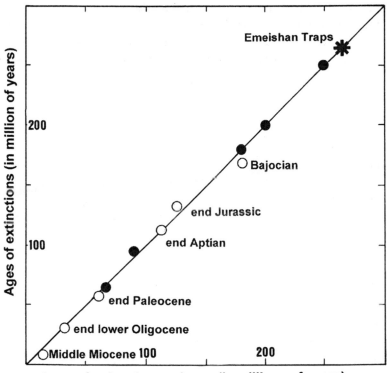

Figure 5.3. Supposed correlation of extinction events and large volcanic provinces. This is redrawn from Courtillot (1999), which in turn was based upon an original figure from Rampino and Stothers (1988). Here it has been modified to show the true correlations (solid spots) and spurious correlations (open circles) where the volcanism occurred at times of no mass extinction. The additional correlation, identified by Hallam and Wignall (1997), of the Emeishan Traps and the end-Guadalupian mass extinction has also been added. The picture that emerges is of a good correlation between 265 and 180 million years ago with only occasional correspondence seen in more recent times.

mainly come about because of dramatic improvements in radiometric dating techniques and the recognition of more giant volcanic provinces than Raup's six. These have shown that, like the Deccan Traps, other flood basalt provinces were erupted extremely rapidly (typically in a million years) and, in many cases coincidentally with mass extinction events (Figure 5.1). A new flood basalt province has also been discovered since 1988, the Emeishan Province of southwest China and this neatly coincides with the also newly discovered end-Guadalupian mass extinction.

Studying Figure 5.1 reveals that the best correspondence between flood basalts and extinctions occurs in the interval beginning with end-Guadalupian event and finishing four events later with the Karoo/Early Toarcian correspondence. Perplexingly, the next great volcanic province formed in the Parana region of Argentina in the Valanginian Stage of the Early Cretaceous, a time marked by some of the lowest extinction rates of the entire fossil record. Some, notably Vincent Courtillot, continue to propound Rampino and Stother's original idea that all flood basalt provinces caused mass extinction events (Courtillot, 1999), but the facts reveal that the correspondence is not one-to-one. This leaves us with a problem.

This problem can be compared with the experiences of a homicide detective called to investigate 10 separate murder cases in which the victim has been shot dead. At seven of the crime scenes he finds the same man standing next to the body but he is not holding a smoking gun (or any other weapon for that matter). This man, a Mr Basalt, refuses to help with enquiries. Would you attempt to prosecute Mr Basalt on the basis of such extraordinary coincidence? It would certainly be tempting, but you would be unlikely to get a conviction. A fruitful line of enquiry would be to see if the circumstances of the murders were similar, and this is the approach many extinction studies are currently pursuing.

Volcanic activity emits various gases of which water vapour, carbon dioxide and sulphur dioxide are the most important. As seen with the Laki eruptions the sulphur dioxide appears to cause the most damaging effects, but these effects were only very short term – the sulphate aerosols are rained out of the atmosphere within a few years. The problem for extinction studies is whether a few years of global cooling can cause long-lasting climatic effects. The thermal heat capacity of the oceans is such that a few years of cooler air temperatures will not change the temperature of the oceans which act as a kind of thermal buffer for the planet. Some proponents of volcanic kill mechanisms have therefore proposed that acid rain may cause the extinctions, but again calculations suggest that this is unlikely. The problem primarily stems from the fact that flood basalt provinces are erupted in discrete pulses with recovery intervals of thousands of years in between. Officer et al. (1987) calculated that the Deccan Traps may have caused lethal acid rain but only by assuming a 10^6 km^3 of basalt were erupted in an initial 10 000-year burst of volcanism, an interval of time that is at least an order of magnitude too brief.

A further problem with sulphur dioxide-caused extinction is the quiet nature of flood basalt eruptions (Wignall, 2001). These occur from elongate fissures, rather than the more normal volcanic cones, and they are the least violent form of volcanic eruption known. As a result they are not capable of propelling gas or dust into the stratosphere, and yet this is a key requirement if the gases are to cause global effects. Once in the stratosphere sulphate aerosols can be circulated around the hemisphere making the cooling effects more than just local. Sulphate aerosols have such a short residence time in the atmosphere that there is insufficient time for them to be circulated within the lower atmosphere (troposphere) before they are rained out. Potentially the giant fissure eruptions, responsible for the largest of flood basalt flows, may have been associated with fire fountains that could have reached up to the stratosphere. Fire fountains have been observed during modern fissure eruptions, for example on Hawaii; they are columns of lava that rise up to several hundred metres above the main eruption. They may have risen several kilometres high above the giant fissure eruptions of the past. Some flood basalt provinces also have large volume flows of pyroclastic rocks, notably the Siberian Traps and Parana Province, which were produced by violent volcanism potentially much more capable of getting gas and dust into the stratosphere. This point has been highlighted for the Siberian Traps in particular, which coincide with the end-Permian mass extinction (Campbell *et al.*, 1992). However, the Siberian Traps were erupted close to the North Pole and stratospheric circulation at such high latitudes is such that it is not capable of circulating gases around the globe, so any cooling effects are likely to have been restricted to the immediate vicinity of the eruptions.

Clearly there are many problems associated with a kill mechanism based on volcanic sulphur dioxide eruptions; however, a mechanism involving volcanic carbon dioxide emissions has more viability. As outlined below, many mass extinction intervals are associated with rapid phases of global warming and oceanic stagnation, two phenomena that are closely linked (Table 5.1). The sudden input of large volumes of volcanic CO_2 into the atmosphere could potentially be a cause of this warming, and it has been commonly invoked (for example by Wignall and Twitchett, 1996). However, once again this potential kill mechanism runs into problems, in this case because of the volumes of gases associated with flood basalt eruptions. Based on measurements from Hawaiian eruptions it is thought that 1 km^3 of basalt would release 5×10^{12} g of carbon (in the form of CO_2) into the atmosphere. Unlike SO_2, CO_2 has a much longer residence time in the atmosphere (thousands of years) before various processes (that include rock weathering and the burial of organic matter in sediments) can remove it, so there

Table 5.1. *Comparison of events contemporaneous with major volcanic
eruptions and major extinction events*

Extinction event	Volcanic province	Climate change	Oceanic change
End Ordovician	None known	Cooling and southern hemisphere glaciation	Improved circulation and oxygenation in deep waters
Frasnian–Famennian	Small volcanic province in the Ukraine	Possibly cooling	Two phases of anoxia
End Guadalupian	Emeishan Traps	Not studied	Anoxia in deep oceans
End Permian	Siberian Traps	Major warming	Prolonged anoxia
End Triassic	Central Atlantic Province	Warming	Anoxia
Early Toarcian	Karoo Traps	Warming	Anoxia
No extinction	Parana Traps	Warming	Anoxia
Cenomanian–Turonian	Caribbean–Columbian	Major warming	Anoxia
End Cretaceous	Deccan Traps	Brief cooling and then warming	Possibly deep ocean anoxia
No extinction	North Atlantic Igneous Province	Brief interval of intense warming	Brief oxygen-poor interval in ocean depths
No extinction	Ethiopian Traps	No effect	No effect
No extinction	Columbia River Basalts	No effect	No effect

is ample time for the gas emitted to diffuse around the world's atmosphere. How much carbon dioxide was erupted per lava flow depends
on the size of individual flows and this is currently very poorly known.
Courtillot (1999) suggests 10 000 km^3 and it is unlikely to be much
greater than this figure. Such giant flows would release 5×10^{16} g
of carbon as CO_2 into the atmosphere, which is clearly a lot of gas,
but it is not that much. In comparison, current burning of fossil fuels
releases a similar amount of CO_2 into the atmosphere every year, year
on year, and we have not yet witnessed a catastrophic global warming, at least not yet. This problem was first identified around 20 years
ago (Toon, 1984), and it is a problem that refuses to go away. For this
reason I have suggested that volcanic CO_2 release is more likely to be
a trigger mechanism, rather than the direct cause of global warming

and extinctions (Wignall, 2001). But the trigger for what? As outlined below, volcanoes are not the only geological carbon source capable of injecting large volumes of CO_2 into the atmosphere in a brief interval; gas hydrates buried at shallow depths beneath continental slope sediments are a further source.

In summary, the link between volcanism and extinction is an intriguing one with the excellent correlation of events providing the most compelling evidence, but the mechanism remains unresolved. So, whilst he is the prime suspect for many crimes, Mr Basalt still walks free.

As a final note, there have been several attempts to unify the meteorite impact and giant volcanism kill mechanism by suggesting that impacts caused the volcanism (Alt et al., 1988). The idea neatly side-steps the problem of finding the impact crater, because it becomes infilled with basalt lava flows, but it has generally been greeted with sceptism by the geological community: melting in the upper mantle is unlikely to generate the vast volumes of magma associated with giant volcanic provinces. The idea also singularly fails to explain the lack of volcanism at Chicxulub which is after all one of the largest known craters on Earth. Unperturbed, proponents of the impact-volcanism model have suggested the volcanism will occur at the opposite side of the world to the impact, the antipodal point, where the earthquake waves refocus. However, once again, the Chicxulub impact did not generate antipodal volcanism; the Deccan Traps are not at this point and, for good measure, the Deccan volcanism was well underway before the meteorite struck.

SEA-LEVEL CHANGE

Regression

Sea levels have fluctuated throughout the Earth's history, with the constantly changing volumes of mid oceanic ridges and continental ice sheets being the driving mechanism behind the oscillations. So is it possible that, at certain times, sea-level change could be lethal to life on Earth? Regression (sea-level fall) is certainly one way of reducing the living space of the shallow marine organisms that form the largest component of the fossil record. This could potentially reduce marine species numbers. A link between sea-level change and some groups has indeed been observed, most notably for the ammonites, an extinct group of marine molluscs distantly related to squid. Thus, shortly before their final extinction in the end-Cretaceous cataclysm, ammonite diversity is

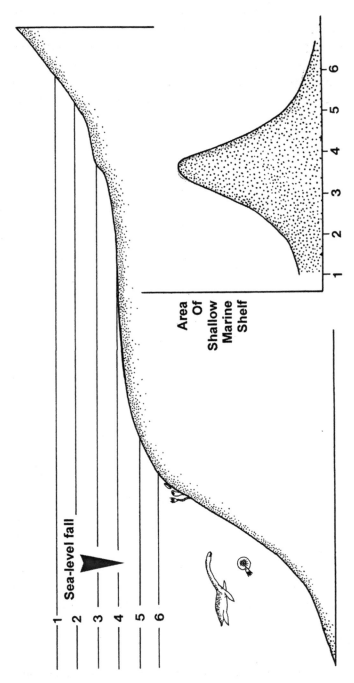

Figure 5.4. Relationship between sea level and area of shallow marine seas. If the original starting position of sea level is very high, as shown here, then sea-level fall produces an increase in shallow marine area.

seen to decline during a regression that just precedes the mass extinction (Ward et al., 1991). But can regression be held responsible for the mass extinctions that have wiped out diverse groups of marine organisms? Newell (1967) was the first to propose just such an idea when he suggested that the magnitudes of regressions and extinctions were correlated. The end-Permian event in particular was held up as an example where the biggest regression of the past 600 million years coincides with the biggest mass extinction (Schopf, 1974). In fact the link between sea level and habitat area is not a direct one because it depends upon the starting position of sea level and continental hypsometry (the distribution of continental elevation). For example, if sea level is very high, as it was throughout much of the Cretaceous (Hallam, 1992), then the area of shallow water is likely to be quite restricted and may actually increase if sea level falls (Wyatt, 1987; Figure 5.4).

The end-Permian regression/mass extinction link, the cornerstone of Newell's hypothesis, has not withstood the test of further investigation. Only in recent years have the detailed sea-level changes of the Permian–Triassic interval been investigated and integrated with evidence for the extinction level. These studies have revealed that, whilst the interval from the Early Permian to the Middle Triassic was marked by generally low sea levels, sea level was actually rising rapidly across the Permian–Triassic boundary. The mass extinction occurs within this interval of transgression (Wignall and Hallam, 1992; Figure 5.5). In fact the low point of sea level occurs at the Middle/Upper Permian boundary suggesting that regression and habitat loss may be more important in explaining the end-Guadalupian mass extinction. Regression may also be an important component of the extinction mechanism for both the end-Ordovician and end-Triassic mass extinctions (Hallam and Wignall, 1999). In the former case, sea-level fall was very rapid and was probably caused by the growth of a continental ice sheet in the southern hemisphere. The origin of the Late Triassic regression is less obvious; glaciation is not known from this interval. Some of the best evidence for sea-level fall during the latest Triassic occurs around the North American–European region raising the possibility that the regression may have been caused by regional uplift immediately prior to the eruption of the Central Atlantic Magmatic Province in this area (Hallam, 1997).

Transgression

Rapid sea-level rise also occurs during several mass extinction intervals, notably the end-Permian event, as already noted, and during

Figure 5.5. Large cliff section at Butterloch in the Italian Dolomites showing the Permian to Triassic transition. A sharp boundary (arrowed) marks a sea-level fall in the Late Permian. The extinction level, marked with an asterisk, occurs in the overlying transgressive sedimentary record.

the Frasnian–Famennian, early Toarcian, Cenomanian–Turonian and end-Cretaceous events. Furthermore, both the end-Ordovician and end-Triassic low points of sea level were immediately followed by intervals of rapid rise (Hallam and Wignall, 1999). Transgressions are not an obvious cause of extinction because they generally increase marine habitat area; however, many transgressions coincide with the development of waters lacking oxygen – anoxic waters – a much more lethal potential kill mechanism, as outlined below.

MARINE ANOXIA

Modern oceans are extremely well ventilated, with ample dissolved oxygen being available at all water depths, right down to the kilometres-deep abyssal depths of the world's great oceans. This remarkable observation is due to a vigorous circulation regime. Most oceanic deep waters are generated in polar shelf seas where cold, dense, oxygen-rich waters form and then sink (due to their density) into the ocean depths. Here they begin a remarkable journey as part of a giant conveyor belt-like circulation system that ultimately returns the waters to the poles as warmer, less dense, surface water. Only in certain, specific locations such as basins partially isolated from the main oceans or in severely-polluted shallow seas are anoxic conditions developed. Thus, it was with some surprise that the first systematic drilling of oceanic sediments by the Deep Sea Drilling Program in the 1970s revealed that there had been short intervals of time when the world's oceans appear to have gone anoxic (Schlanger and Jenkyns, 1976). These 'oceanic anoxic events', or OAEs, were identified in Cretaceous rocks, with the most widespread occurring at the Cenomanian–Turonian boundary. A little under a decade later this interval was identified as a time of mass extinction, and thus the first link with global marine anoxia and extinction was established.

Perhaps one of the most remarkable discoveries of the past decade is the fact that nearly all mass extinctions, not just the Cenomanian–Turonian one, correspond with intervals of widespread marine anoxia (Table 5.1). Ocean floor older than the late Jurassic has all been subducted into the Earth's mantle, making it difficult to determine if these pre-Cretaceous anoxic events were truly oceanic in their extent. However, sufficient clues remain to suggest that they probably were. Some oceanic sediments get scraped-off the oceanic crust on which they sit and become part of the mountain belts that override subduction zones. These sediments form part of accretionary prisms and large areas of Japan are formed of such material. Rocks here have revealed that the anoxic events of the Early Triassic and Early Toarcian are likely to have extended into the ocean depths (Isozaki, 1997).

Chemical records of limestones provide a second line of evidence for the extent of marine anoxia. Anoxic conditions favour the preservation of organic matter in sediments with the result that many anoxic sediments are black shales; the black colour coming from their

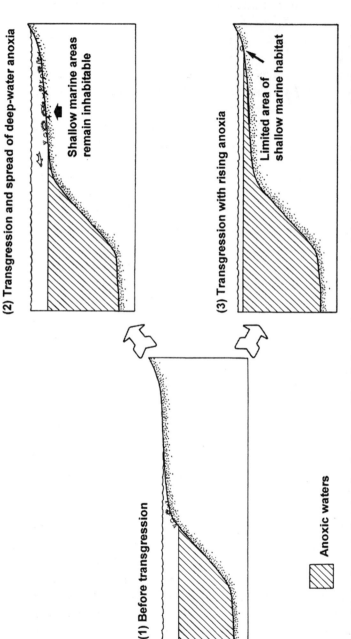

Figure 5.6. A schematic diagram showing the possible relationship between sea-level rise, anoxia and the area of shallow marine deposition. In situation (2) the transgression introduces deep water, anoxic conditions into deep continental shelf locations. Such a scenario may have happened during the minor Cenomanian–Turonian boundary extinction. In situation (3) the transgression coincides with a rise in the level of anoxic bottom waters with the result that shallow marine habitat areas became severely restricted. Such a scenario appears to have happened during the end-Permian mass extinction, the most severe of the fossil record.

high organic carbon content. This carbon is derived from plant matter, either plant material swept into the oceans from land or from marine plankton. In the natural world carbon comes in two stable forms, ^{12}C and the much rarer ^{13}C. There is also a third isotope, ^{14}C, but this is so highly radioactive that it decays rapidly and is not preserved in the geological record. The formation of organic carbon involves the preferential use of the light isotope ^{12}C, with the result that it is said to be isotopically light. This is a very important characteristic because during oceanic anoxic events so much organic carbon is preserved and buried in sediments that the remaining carbon in the world's oceans and atmosphere starts to become enriched in ^{13}C; it is said to become isotopically heavier. Fortunately for geologists the isotopic composition of the world's oceans is preserved in limestones. Carbon is present in limestones as part of the carbonate ion, and the formation of limestones essentially records the isotopic composition of the waters from which it formed; in technical parlance there is said to be little fractionation. Thus, limestones formed during oceanic anoxic events record a change to heavier (more ^{13}C rich) carbon values. This is precisely what is observed during many such events and for some of the Palaeozoic events, when we have no preserved rock record from ocean sediments or accreted terranes, it is the main line of evidence for OAEs. For example, the F–F mass extinction of the Late Devonian coincides with the development of two widespread organic-rich beds, called the Kellwasser Horizons from the name given to them in German localities. Both the Kellwasser Horizons coincide with a marked shift to more ^{13}C-rich oceans, as recorded by limestones, suggesting that they record global oceanic anoxic events.

The connection between marine anoxia and marine mass extinction is well established, but the proximate kill mechanism is a subject of debate. As noted above, the development of anoxia often coincides with transgression. This may mark the development of deep-water anoxia in areas previously characterized by shallower more oxygenated conditions. This need not necessarily cause extinction as long as the shallow marine organisms can migrate and the area of shallow marine habitat does not decrease significantly (Figure 5.6). As noted above, this last factor depends upon the original starting level and hypsometry. However, some OAEs, notably the severe and prolonged one associated with the end-Permian mass extinction, not only record the expansion of the area of deep-water anoxia but also the development of anoxia in exceptionally shallow water depths. Some earliest Triassic localities record the development of oxygen-poor waters in exceptionally shallow waters, perhaps as little as 10 m deep. In such a situation marine habitat area becomes of very limited extent indeed. The development of anoxia in the world's oceans also has a profound affect on oceanic nutrient supply and it is probable that many of the planktonic extinctions seen during OAEs are attributable to dramatic changes in nutrient dynamics (Hallam and Wignall, 1997).

The ultimate cause of marine anoxia is intensely debated and it is possible to write a book about this subject without coming to a definitive answer (see Wignall, 1994). Many OAEs (and mass extinctions) appear to coincide with phases of global warming (see below), and the effect such changes have on oceanic circulation. The key factor may be a reduction in the rate of deep-water generation in polar regions as the poles warm up, with the result that oceanic circulation begins to slow down or stagnate (Wignall and Twitchett, 1996). Some geologists have suggested that global warming may cause a shift in the site of formation of deep ocean waters from the poles to the tropics. Thus, it is argued, the waters in shallow shelf seas will become warm and saline (due to the effects of evaporation) with the result that their density increases allowing them to sink into the deep oceans. It is not immediately apparent why such warm saline bottom water (WSBW), as it is called, should be anoxic, although it is important to note that the solubility of oxygen in water decreases rapidly with temperature increase. A telling argument against WSBW-generated oceanic anoxia comes from the Mediterranean, the only ocean today with a circulation system driven by surface water evaporation. As a result Mediterranean deep waters are warm and saline but they are also O-rich and well ventilated; the antithesis of an oceanic anoxic event.

GLOBAL WARMING

In the previous sections it has been proposed that global warming is a recurrent theme during many mass extinction events. The evidence for these changes comes from a broad range of palaeoclimatic sources including a battery of new techniques not available a few years ago. One of these, the 'stomatal index', has proved particularly powerful. Stomata are small holes on the surface of a leaf that allows it to take in carbon dioxide and photosynthesize. They also allow water vapour to escape, an unwanted side effect; therefore a leaf's stomatal density is a balance between the needs of CO_2 uptake and the minimizing of water loss. Fortunately for leaves, as the CO_2 content of the atmosphere increases so the number of stomata required decreases. These changes are what the stomatal index quantifies, with lower values signifying more CO_2-rich conditions and thus a more greenhouse-like climate. Stomatal evidence from the earliest moments of the Triassic and the Jurassic indicate dramatic increases of atmospheric CO_2 levels (McElwain *et al.* 1999; Retallack, 2001), and thus imply global warming during the immediately preceding mass extinction intervals. Further stomatal index studies will no doubt help clarify temperature changes during other crises.

Global warming *per se* is unlikely to lead to the direct extinction of many marine organisms, although warmer conditions heighten metabolic activity and thus increase O requirements and, as seen in the previous section, warming, marine anoxia and transgression, often go hand-in-hand. On land the principal detrimental effect of warming is seen in the decline of cold, high latitude habitats. This occurred in extreme form during the end-Permian mass extinction which saw the total eradication of high latitude forests including southern hemisphere vegetation dominated by *Glossopteris* trees (Figure 5.7; see also Chapter 3). The Early Triassic at all latitudes is remarkable for a flora lacking plants taller than a metre (a world without trees!) and for having polar climates more typical of that encountered in the south of France or California today (Retallack, 1999). Cold-adapted plants, able to tolerate freezing conditions, had no place in the Early Triassic world.

Mechanisms for achieving such an extreme greenhouse climate hinge on the postulated source of CO_2. As noted above, the deadly triumvirate of flood basalt volcanism, marine anoxia and global warming are often to be found during mass extinction intervals, and any kill mechanism must try and link these three phenomena (Kerr, 1998). Basalt volcanism certainly releases CO_2 into the atmosphere but, as noted above, even the largest such eruption may not trigger a substantial temperature increase. It may provide a trigger though. Recent investigations of sediments on the continental shelf have made the extraordinary discovery that vast volumes of methane are trapped, within ice crystals, beneath the seafloor. This material, called gas hydrate, is in a delicate equilibrium; only a minor temperature change or a slight decrease of pressure would be sufficient for the ice to melt and the methane to escape to the atmosphere. Methane is an extremely effective greenhouse gas although it rapidly oxidizes to CO_2, a slightly less effective although nonetheless potent greenhouse gas. Thus, slight global warming, triggered by massive volcanism, may be sufficient to start a runaway greenhouse climate in which methane is released to the atmosphere causing further warming and further methane release (Figure 5.8). An additional nasty positive feedback mechanism comes

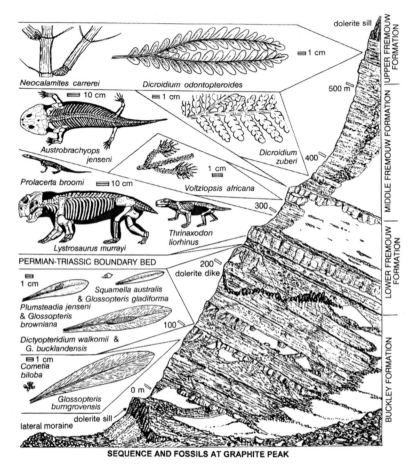

Figure 5.7. Sketch of the Graphite Peak cliff section (central Transantarctic Mountains, Antarctica) and associated fossils. The Permian–Triassic boundary 250 m above the base of the section marks the loss of a flora dominated by glossopterid leaves and the appearance a fossil assemblage that includes *Lystrosaurus*, a genus found throughout the world in the immediate aftermath of the end-Permian extinction. Reproduced from Retallack and Krull (1999), with the permission of the authors and the Geological Society of Australia.

from the release of CO_2 in solution as the oceans warm up. If sulphate aerosols from flood basalt volcanism are temporally able to acidify the surface waters of the oceans then yet more carbon dioxide will be released. The end result of all this global warming appears to be oceanic stagnation and widespread marine anoxia (see above) and marine mass extinction. On land the loss of cold climates is catastrophic for high

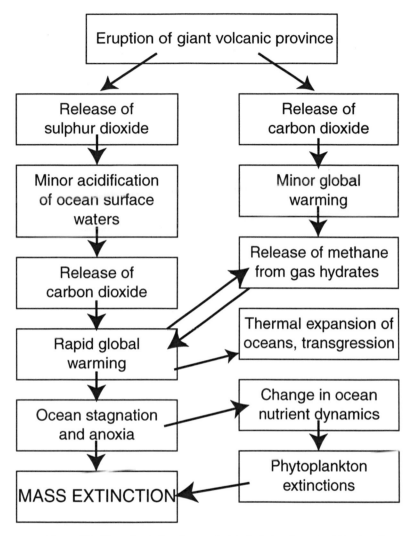

Figure 5.8. Flow chart showing postulated chain of events following the eruption of a giant volcanic province (based on models proposed by Wignall and Twitchett (1996), Kerr (1998) and Wignall (2001)).

latitude ecosystems although lower latitude plants and animals should suffer less. Their demise may be linked with the climatic extremes generated by a supergreenhouse climate, particularly in the centres of continents, where temperatures may have become lethally hot. The chain of causes and events listed in Figure 5.8 neatly explains many phenomena seen during several mass extinction events, and it also provides a mechanism to ultimately halt the catastrophic chain of events. Marine anoxia causes enhanced burial of organic carbon and so draws down atmospheric carbon dioxide (remember that organic carbon is a product of photosynthesis that uses atmospheric CO_2), although this is only a gradual process not a quick-fix solution to the planet's problems.

Support for the key role played by gas hydrates comes, once again, from the carbon isotope record of limestones. The onset of several mass extinctions (notably the end-Permian, end-Triassic and early Toarcian events) is marked by a brief shift to extremely ^{12}C-rich limestones followed by a return to less rich values. Gas hydrate methane is also very ^{12}C-rich and it is obviously tempting to see this trend as reflecting a sudden release of methane into the atmosphere, with the subsequent trend reflecting the burial of light organic carbon during the oceanic anoxic event. As a final word of caution, it should be noted that gas hydrates are currently a very trendy subject in geology and there is a tendency to attribute every carbon isotope event to release of this volatile material. However, gas hydrate release is not the only way to cause limestones to become isotopically light. The catastrophic death of most of the world's plants would achieve the same result and, as noted below, such a doomsday scenario is not at all far-fetched for some mass extinction events.

GLOBAL COOLING

Global warming figures in several models for mass extinction, but the reverse trend, global cooling has also been blamed for some events. For example, the cooling effect of volcanic aerosols has been suggested as a contributory cause of the end-Permian mass extinction (Campbell *et al.*, 1992). However, as noted above the short duration of this effect, a few years at most, is unlikely to change the Earth's climate and there is no supporting evidence from the geological record for any latest Permian cooling. More substantial evidence for cooling occurs in the immediate aftermath of the F–F mass extinction in the Late Devonian. Most of this evidence comes from fossils, in particular the proliferation of sponges that construct their skeletons from silica. Such siliceous sponges are typical of deep, cold water habitats but after the late Devonian extinction they are found in shallower water at low palaeolatitudes. It has also been argued that the preferential loss of species from warmer water locations during this extinction event is good evidence in favour of a death-by-cooling extinction model (Copper, 1986). However, a word of caution, it should be noted that our knowledge of cool and cold water marine species in the Late Devonian is very poor because the fossil record from high latitude locations of this age is meagre. Thus, it is not really possible to say whether the extinctions were any more severe in tropical latitudes than they were in higher latitudes.

Perhaps the most convincing link between global cooling and extinction occurred during the first of the great extinction events at the close of the Ordovician. This interval witnessed several, no doubt related, phenomena including a rapid draw-down of sea level, and the development of extensive continental ice sheets. Paradoxically, this ice was developed in some of the hottest places on Earth today, the Sahara desert and the Amazon basin, but 430 Ma ago these regions formed part of a large continent that lay over the South Pole. This was the first of three major ice ages that have occurred in the past 600 Ma, with the end-Ordovician being the briefest with a duration of around a million years. The second ice age spanned roughly 80 Ma from the Early Carboniferous to the Early Permian, whilst we are currently living through the third ice age which began several million years ago. However, of these three chilly intervals only the first event appears to have caused a mass extinction. So, why was the end-Ordovician glaciation so lethal?

The answer may partly lie in the rapidity of the ice age's onset. Ordovician glaciation appears to have both started and finished abruptly with consequent rapid changes in environments (Hallam and Wignall, 1997). The associated sea-level fall, caused by the growth of continental ice, further caused large areas of shallow marine seas in North America to become newly exposed land, with consequent habitat loss for large numbers of marine invertebrates. Thus, the extinctions of many species are probably attributable to sea-level fall (the proximate cause) although glaciation is the driving mechanism (the ultimate cause). The preferential loss of tropical species suggests that cooling was also an important contributing factor. However, when viewed in detail the timing of end-Ordovician extinctions resolves itself into two distinct extinction phases, and points to a more complex extinction mechanism. The first of these phases coincides with glaciation and sea-level fall, as already mentioned, but the second of the phases occurred at the end of glaciation when sea-levels rose once again and the interiors of continents were once again flooded. Many of the victims of the second extinction were deep water species that had escaped the first crisis. Their demise seems to once again be related to that nemesis of marine diversity – the development of anoxic waters at a time of extinction.

STRANGELOVE OCEANS

Since Sepkoski's original compilation of extinction rates in the fossil record many have wondered whether the Big Five mass extinctions are somehow different in kind from the general run of extinctions. When extinction rates are ranked in order of magnitude, the mass extinctions seem to simply represent the end members of a gradual decreasing continuum and so, statistically at least, they do not form a separate group. However, they may differ in one key respect: the Big Five mass extinction events appear to record the only intervals when there has been a near-total shut-down of marine primary productivity. In other words, they are times when the tiny, single-celled photosynthesizing algae that form the base of the marine food chain disappeared – a catastrophic occurrence that goes a long way to explaining the extinctions of other marine groups. This is the 'Strangelove ocean' scenario first proposed to explain the oceanic situation in the immediate aftermath of the meteorite impact at Chicxulub (Hsu and McKenzie, 1985). The evidence in this case is based upon both palaeontological and carbon isotope data. The end Cretaceous saw the extinction of most coccolithophores, a group of phytoplankton that secrete tiny plates of calcite (thus ensuring that they have a good fossil record), and many planktonic foraminifera. This latter group of millimetre-sized protists are not phytoplankton, but rather they graze on them, and so their fortunes are intimately linked. The demise of these planktonic groups also coincides with a change in the carbon isotope record of shallow marine carbonates, which show an increase in the relative abundance of ^{12}C. This change is interpreted as a record of the loss of the ^{12}C-rich organic matter of phytoplankton from the surface waters with a resultant enrichment of surface waters in the light isotope.

Other mass extinctions also show the abrupt demise of planktonic groups. Up until the Cretaceous the radiolarians were the most common microplankton in the oceans. These are another group of protists that graze on phytoplankton. They construct complex and often beautiful skeletons from a delicate latticework of silica. During both the end-Permian and end-Triassic mass extinctions the radiolarians underwent catastrophic extinctions, with the former event almost causing their total extinction. This is good evidence for a collapse of marine primary productivity, without being direct evidence. The mass extinction at the end of the Permian was so severe that it changed the nature of deep ocean sedimentation for millions of years. Generally, deepwater sediment accumulation in sites far removed from the reach of

Figure 5.9. Tentaculitoids in a deep water limestone of Frasnian (Late Devonian) age from the Holy Cross Mountains of central Poland. Such tiny cones were prolifically abundant in the surface waters of Late Devonian oceans until their abrupt extinction at the Frasnian–Famennian boundary. Photograph taken by Dave Bond using a scanning electron microscope. Scale bar, 100 μm.

land-derived sediments, such as in the middle of a large ocean, is often dominated by the rain of planktonic skeletons, such as radiolarians; radiolarian chert is the resultant rock type. Following the end-Permian extinction, chert formation in the Triassic oceans ceased for up to 10 Ma until the radiolarians gradually pulled back from the brink of total extinction.

If we consider the older mass extinctions, the main planktonic groups become ever more strange and unfamiliar. Thus, in the Late Devonian the tentaculitoids were the dominant group of fossil plankton and many deep-water limestones from this time are almost entirely composed of these cones (Figure 5.9), in much the same way that deep-water limestones today are often composed of planktonic foraminifera. Despite their evident success, the tentaculitoids disappeared abruptly during the F–F mass extinction. During the oldest mass extinction at the end of the Ordovician, water column dwellers appear to have been particularly unlucky with nearly all planktonic groups disappearing at this time (Hallam and Wignall, 1997).

Thus, a holocaust in the water column is the unifying feature of true mass extinction events, but this does not necessarily imply a single cause. As this review has hopefully made clear there are numerous viable and competing extinction mechanisms. Somewhat ironically, the intense interest shown by geologists in the origin of mass extinctions was triggered by the suggestion that a meteorite impact caused the end-Cretaceous mass extinction and yet this kill mechanism has not been found to provide a general mechanism for mass extinctions. The last day of the Cretaceous was both dreadful and unique. Other mass extinctions seem more explicable in terms of Earth-bound processes and events, with only massive volcanism coming close to providing a single unifying cause, and even here the correlation is better established than the causation.

FURTHER READING

Albritton, C. C. Jr *Catastrophic Episodes in Earth History*. London: Chapman and Hall, London, 1989.

Archibald, J. D. *Dinosaur Extinction and the End of an Era: What the Fossils Say*. New York: Columbia University Press. 1996.

Lavers, C. *Why Elephants Have Big Ears: Understanding Patterns of Life on Earth*. London: Phoenix Paperbacks. 2001.

Officer, C. and Page, J. *The Great Dinosaur Extinction Controversy*. Reading: Addison-Wesley Publishing Company. 1996.

Powell, J. L. *Night Comes to the Cretaceous: Comets, Craters, Controversy, and the Last Days of the Dinosaurs*. San Diego: Harcourt Brace and Co. 1998.

REFERENCES

Alt, D., Sears, J. M. and Hyndman, D. W., 1988. Terrestrial maria; the origins of large basalt plateaus, hotspot tracks, and spreading ridges. *Journal of Geology* **96**: 647–662.

Alvarez, W. 1997. *T. rex and the Crater of Doom*. Princeton: Princeton University Press.

Alvarez, L. W., Alvarez, F., Asaro, F. and Michel, H. V., 1980. Extraterrestrial cause for the Cretaceous–Tertiary extinction. *Science* **208**: 1095–1108.

Becker, L., Poreda, R. J., Hunt, A. G., Bunch, T. E. and Rampino, M., 2001. Impact event at the Permian–Triassic boundary: evidence from extraterrestrial noble gases in fullerenes. *Science* **291**: 1530–1533.

Bohor, B. F., Foord, E. E., Modreski, P. J. and Triplehorn, D. M., 1984. Mineralogic evidence for an impact event at the Cretaceous–Tertiary boundary. *Science* **224**: 867–869.

Campbell, I. H., Czamanske, G. K., Fedorenko, V. A., Hill, R. A. and Stepanov, V., 1992. Synchronism of the Siberian Traps and the Permian–Triassic boundary. *Science* **258**: 1760–1763.

Carlisle, D. B. 1995. *Dinosaurs, Diamonds, and Things from Outer Space. The Great Extinction*. Stanford: Stanford University Press.

Copper, P., 1986. Frasnian/Famennian mass extinction and cold-water oceans. *Geology* **14**: 835–839.

Courtillot, V. 1999. *Evolutionary Catastrophes: The Science of Mass Extinction.* Cambridge: Cambridge University Press.

Hallam, A., 1992. *Phanerozoic Sea-level Changes.* New York: Columbia University Press.

1997. Estimates of the amount and rate of sea-level change across the Rhaetian–Hettangian and Pliensbachian–Toarcian boundaries (latest Triassic to early Jurassic). *Journal of the Geological Society of London* **154**: 773–779.

Hallam, A. and Wignall, P. B. 1997. *Mass Extinctions and their Aftermath.* Oxford: Oxford University Press.

1999. Mass extinctions and sea-level changes. *Earth-Science Reviews* **48**: 217–250.

Hildebrandt, A. R., Pilkington, M., Cannors, M., Ortiz-Alema, C. and Chavez, R. E., 1995. Size and structure of the Chicxulub crater revealed by horizontal gravity gradients and cenotes. *Nature* **376**: 415–417.

Hsu, K. J. and McKenzie, J. A., 1985. A 'Strangelove' ocean in the earliest Tertiary. *American Geophysical Union Monograph* **32**: 487–492.

Isozaki, Y., 1997. Permo-Triassic boundary superanoxia and stratified superocean: records from lost deep sea. *Science* **276**: 235–238.

Kaiho, K., Kajiwara, Y., Nakano, T. *et al.*, 2001. End-Permian catastrophe by a bolide impact: evidence of a gigantic release of sulphur from the mantle. *Geology* **29**: 815–818.

Kerr, A. C., 1998. Oceanic plateau formation: a cause of mass extinction and black shale deposition around the Cenomanian–Turonian boundary. *Journal of the Geological Society of London* **155**: 619–626.

McElwain J. C., Beerling, D. J. and Woodward, F. I., 1999. Fossil plants and global warming at the Triassic–Jurassic boundary. *Science* **285**: 1386–1390.

McGhee, G. R. Jr, 2002. The 'multiple impacts hypothesis' for mass extinction: a comparison of the Late Devonian and the late Eocene. *Palaeogeography, Palaeoclimatology, Palaeoecology* **176**: 47–58.

Newell, N. D., 1967. Revolutions in the history of life. *Geological Society of America Special Paper* **89**: 63–91.

Officer, C. B., Hallam, A., Drake, C. L. and Devine, J. D., 1987. Late Cretaceous and paroxysmal Cretaceous/Tertiary extinctions. *Nature* **326**: 143–149.

Rampino, M. R. and Stothers, R. B., 1988. Flood basalt volcanism during the past 250 million years. *Science* **241**: 663–668.

Raup, D. M. 1991. *Extinction, Bad Luck or Bad Genes?* New York: W. W. Norton & Co.

Retallack, G., 1999. Postapocalyptic greenhouse paleoclimate revealed by earliest Triassic paleosols in the Sydney Basin, Australia. *Bulletin of the Geological Society of America* **111**: 52–70.

Retallack, G. J., 2001. A 300-million-year record of atmospheric carbon dioxide from fossil plant cuticles. *Nature* **411**: 287–290.

Retallack, G. J. and Krull, E. S., 1999. Landscape ecological shift at the Permian–Triassic boundary in Australia. *Australian Journal of Earth Sciences* **46**: 785–812.

Retallack, G. J., Seyedolali, A., Krull, E. S., Holser, W. T., Ambers, C. A. and Kyte, F. T., 1998. Search for evidence of impact at the Permian–Triassic boundary in Antarctica and Australia. *Geology* **26**: 979–982.

Schlanger, S. O. and Jenkyns, H. C., 1976. Cretaceous oceanic anoxic events: causes and consequences. *Geologie en Mijnbouw* **55**: 179–184.

Schopf, T. J. M., 1974. Permo-Triassic extinctions: relation to seafloor spreading. *Journal of Geology* **82**: 129–143.

Sepkoski, J. J. Jr, 1994. Extinction and the fossil record. *Geotimes* **March**: 15–17.

Stanley, S. M. and Yang X., 1994. A double mass extinction at the end of the Paleozoic era. *Science* **266**: 1340–1344.

Toon, O. B., 1984. Sudden changes in atmospheric composition and climate. In: Holland, H. D. and Trendall, A. F. (Eds.), *Patterns of Change in Earth Evolution*. Berlin: Springer-Verlag, pp. 41–61.

Ward, P. D., Kennedy, W. J., MacLeod, K. J. and Mount, J. F., 1991. Ammonoid and inoceramid bivalve extinction patterns in the Cretaceous/Tertiary boundary sections of the Biscay region (southwestern France, northern Spain). *Geology* **19**: 1181–1184.

Wignall, P. B. 1994. *Black Shales*. Oxford: Oxford University Press.

   2001. Large igneous provinces and mass extinctions. *Earth-Science Reviews* **53**: 1–33.

Wignall, P. B. and Hallam, A., 1992. Anoxia as a cause of the Permian/Triassic extinction: facies evidence from northern Italy and the western United States. *Palaeogeography, Palaeoclimatology, Palaeoecology* **93**, 21–46.

Wignall, P. B. and Twitchett, R. J., 1996. Oceanic anoxia and the end-Permian mass extinction. *Science* **272**: 1155–1158.

Wyatt, A., 1987. Shallow water areas in space and time. *Journal of the Geological Society of London* **144**: 115–120.

DAVID JABLONSKI

Department of Geophysical Sciences, University of Chicago, 5734 South Ellis Avenue, USA

6

The evolutionary role of mass extinctions: disaster, recovery and something in-between

INTRODUCTION

The fossil record is punctuated by extinction events at all scales, from the loss of one or two fish species with the drying of a lake, to the wholesale disappearance of dinosaurs or ammonites at the end of Cretaceous period 65 million years (Ma) ago. The handful of events that are global in scale and affect a broad spectrum of organisms are termed mass extinctions. Although most research has centred on the causes of mass extinctions (Chapter 5), there has also been a growing appreciation of the evolutionary consequences of mass extinctions. The evolutionary bursts that follow mass extinctions may be as important as the extinction events themselves in setting the tone of the post-extinction world, as new or previously obscure lineages take advantage of the opportunities opened up by the demise of dominant groups. The most familiar example of this came at the end of the Cretaceous Period. Dinosaurs and mammals originated almost simultaneously in the Triassic Period about 225 Ma ago, and the dinosaurs dominated terrestrial ecosystems for over 120 Ma while the mammals lived in the nooks and crannies of the dinosaurs' world. The dinosaurs became extinct 65 Ma ago, along with many other lineages on land and in the sea, bringing the Mesozoic Era to a close, and within the first 10 or 15 Ma of the Cenozoic, a rich mammalian fauna diversified on the land, bats took to the sky and whales to the sea.

This picture of the triumphant rise of the mammals, triggered by mass extinction rather than the supposed adaptive superiority of the mammalian line, should not be taken as the only possible post-extinction fate of surviving clades (rigorously defined evolutionary

Extinctions in the History of Life, ed. Paul D. Taylor.
Published by Cambridge University Press. © Cambridge University Press 2004.

groups, which include all of, and only, the descendants of a given ancestor). A closer look at the fossil record of the recovery intervals following mass extinctions shows that survival alone does not guarantee evolutionary success: contrary to a popular television series, not all survivors are winners. I will start with a discussion of the rules of survival during mass extinctions, go on to the post-extinction drama, and end with a few words on the implications of the fossil record for present-day biodiversity.

WHO SURVIVES?

The Big Five

Taylor (Chapter 1) has already discussed the general distribution of extinction events through the geological record. When plotted through time, the history of life shows many ups and downs, some modest and some very dramatic. Because the fossil record varies in completeness through time, some of the smaller apparent extinctions and apparent pulses in diversification are likely to be artefacts of, or at least accentuated by, uneven preservation. However, the Big Five extinction events, where more than 50 per cent of marine genera are lost from the fossil record, appear to be genuine episodes of extinction, although even here the precise details of timing and intensity are subject to distortion by the fossil record. For example, sea-level drops occur at roughly the same time as the end-Ordovician, end-Permian and end-Cretaceous mass extinctions. These tend to reduce the area or volume of marine sediment available for sampling during the extinction interval; this geological effect will in turn exaggerate the size of apparent extinction (because smaller outcrop areas yield fewer species even if the species were alive). Depending on their exact size and distribution, sampling gaps can make an extinction look artificially abrupt, artificially gradual or artificially stepped.

Despite such imperfections of the geological record, many lines of evidence support the reality of the Big Five: they are seen in global compilations of the last occurrences of taxa (e.g. Sepkoski, 1997); in the loss of abundant and widely distributed, diverse clades such as the ammonites and dinosaurs whose absences from the post-extinction interval are convincing even when sampling is more limited (if there were Cenozoic ammonites or dinosaurs we would have heard about it); and in detailed study of individual outcrops and deep-sea cores in many regions of the world. Increasingly sophisticated mathematical models for the combination of extinction, origination and preservation that best accounts for the history of diversity through time (Foote, 2003) also tend to verify the major extinction events. The Late Devonian extinction remains problematic when studied in global databases because it is smeared out over a broader time interval, for preservational or other reasons (Bambach *et al.*, 2002; Foote, 2003), but outcrop and regional studies do indicate some form of major upheaval in both land and sea communities during this time (McGhee, 1996; Balinski *et al.*, 2002; Copper, 2002; Racki and House, 2002).

Extinction selectivity

The study of extinction selectivity – patterns of extinction and survival among evolutionary lineages or according to biological characteristics of the organisms – has only just begun. We especially need more analyses that compare selectivities among the major extinction events, and across extinction events of different magnitudes. Three basic models for the evolutionary effects of mass extinctions hinge on these selectivities:

(1)   Mass extinctions could simply be intensifications of normal, 'background' extinction. The most extinction-prone clades might be hit especially hard, the most extinction-resistant ones might do relatively well, and the broader outlines of life might go on unchanged. Under this regime, mass extinctions would tend to remove declining or newly established clades and thus reinforce the status quo for incumbents (groups that are well established in their ecological roles and thus pre-empt others from those lifestyles). The long-term history of clades could be predicted by, or extrapolated from, their comparative success during the relatively quiet times of less intense turnover between the major mass extinctions.

(2)   Mass extinctions could be entirely random. Survivorship might be based simply on the luck of the draw, with no selectivity whatsoever. This would really shake up the biosphere, and would virtually demolish our ability to make sense of the patterns of extinction and survival seen in the fossil record, except for the simplest statistical statements.

(3)   Mass extinctions could be selective, but follow different rules than the quieter times that make up the vast majority of the geological past. From an evolutionary point of view, this is in some ways equivalent to the random model, because the traits

contributing to success during quieter times need have little connection to the ones that promote survival during the extinction
event. But it differs in one crucial sense: we can detect selectivity
patterns, which allow us to understand at least some of the winners and losers in an extinction event, and perhaps even make
some predictions about future extinctions.

Although more work comparing selectivities is urgently needed,
evidence is accumulating in favour of model 3, especially for the most
extensively studied of the Big Five extinctions, the end-Cretaceous event
65 Ma ago (Jablonski, 1995; Gould, 2002: chapter 12). For example, during quieter times, clades that contain many species tend to be more
extinction-resistant than species-poor clades. This makes sense: the big
clades are buffered against the loss of a few or even half of their
species, but small clades have few species to spare, and with a little bad
luck they can disappear completely. During the end-Cretaceous extinction, however, species-richness is no guarantee of survival for molluscan
clades (I should note that the sparsely sampled dinosaurs, pterosaurs
and other vertebrates are much less suitable for statistical analyses of
victims and survivors than the less dramatic but far more abundant
marine invertebrate fossils such as clams, snails and sea urchins). This
can be seen in North America, and, contrary to Hallam and Wignall
(1997), in Europe, north Africa, and any other region where the data
are sufficient to perform such an analysis (see Jablonski, 1989, 1995).
Species-richness also appears to confer no advantage for other groups
at this time or during the other four of the Big Five extinctions (see
Jablonski, 1995; Smith and Jeffrey, 1998; but see Erwin, 1989 for an
exception). High local abundances, broad depth distributions, broad
geographic ranges of constituent species, detritus-feeding life habits,
and small body sizes, all of which appear to promote clade survivorship of marine invertebrates during quieter times, were also unimportant during the end-Cretaceous event (Hansen et al., 1993; Jablonski and
Raup, 1995; Harries, 1999; Lockwood, 2003).

Despite this impressive mismatch between extinction selectivity during mass extinctions and quieter times, survivorship is not
strictly random at the end of the Cretaceous. Disproportionate losses
occur for the subset of phytoplankton groups (single-celled photosyn-
thesizers that float in the upper layers of the open ocean) that lack
a dormant stage that can settle onto the seafloor, for corals bearing
symbiotic algae, and the largest and morphologically most complex of

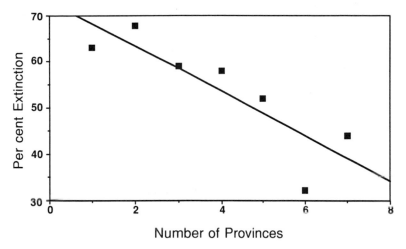

Figure 6.1. Widespread genera of marine bivalves suffer less severe
extinction during the end-Cretaceous mass extinction than
geographically restricted genera. Geographic distribution is measured in
terms of the number of biogeographic provinces in which the genera
occur. Inoceramid bivalves, which evidently became extinct before the
end of the Cretaceous, are omitted from this analysis, as are the bizarre
rudist bivalves that were primarily inhabitants of the shallowest tropical
waters. From Jablonski and Raup (1995).

the planktonic foraminifera (shelled protozoans that also float in the
open ocean) (Kitchell *et al.*, 1986; Norris, 1991, 1992; Racki, 1999; Rosen,
2000). One very general pattern has emerged: broad geographic distribution at the clade level, regardless of the ranges of individual species
within the clade, clearly enhances survival during mass extinctions.
This can be seen not only in end-Cretaceous molluscs (Figure 6.1), but
in many other groups and events (Jablonski, 1995; Harper and Rong,
2001). The geographic ranges of the individual species are no longer
important – the rules have shifted – and survival now depends on the
geographic range of the entire clade, that is, the number of ocean basin
or continent shelves they occupied.

These contrasts in selectivity between mass extinctions and times
of low extinction intensity help to explain why those rare events are so
important in the history of life, even though they account for less than
five per cent of the total species extinction in the geological record
(Raup, 1991). Clades or adaptations can be lost *not* because they are
poorly adapted to the pre- (or post-) extinction settings, but because

they happen to lack features such as broad geographic deployment that favour survival during the extinction bottleneck. These relatively brief and intense episodes of what Raup (1994) has called 'nonconstructive selectivity' – differences in clade survival that do not promote the long-term adaptation of the biota – can re-channel evolution by knocking out dominant groups and creating a wealth of open ecological opportunities for the survivors. The consequences can be as dramatic as the exuberant post-Cretaceous diversification of the mammals 65 Ma ago and, 120 Ma before that, the great Jurassic diversification of the dinosaurs after the Late Triassic elimination of the previously dominant rauisuchid and other vertebrates (Benton, 1996).

THE COMPLEXITIES OF RECOVERY

The role of extinction in creating evolutionary opportunities is now widely accepted. In evolution as in politics, incumbents have an enormous advantage, and the major extinction events promote diversification of survivors, not only in terms of numbers of species but also in terms of morphological or ecological variety. But this is an oversimplification that obscures some of the real complexities of recovery episodes. For example, it is not widely appreciated that birds diversified in the early Cenozoic along with the mammals (see for example Feduccia, 1999, 2003; Dyke, 2001), and for a time the top predators were large flightless birds (Witmer and Rose, 1991; see also Plotnick and Baumiller, 2000: p. 311), presumably playing a role similar to that vacated by carnivorous dinosaurs (although the biggest birds were over 2 m (6 or 7 feet) tall, and so never reached the dimensions of a *Tyrannosaurus rex*). Mammals were thus probably still the hunted and not the hunters until the giant birds in turn fell by the wayside in the mid-Cenozoic (on most continents), although no one has tested whether the disappearance of the avian predators was the result of competition with up-and-coming mammal clades, climate change, or some other factor. At least one clade of large birds, probably separately derived from a small-bodied lineage, evolved in South America and reached the southern USA during the Pliocene, and only became extinct within the last 2 Ma (Figure 6.2). This example and a growing list of others show that post-extinction dynamics involve more than a simple relay, where the dominant role passes smoothly from one major group to another. These post-extinction recoveries, then, are as much a key to understanding the evolutionary roles of mass extinctions as the extinction events themselves.

Figure 6.2. Large flightless birds evolved to become important carnivores in the early Cenozoic after the extinction of the dinosaurs. This phorusrhacoid, one of the last of the line (*Andalgornis*, from the Pliocene of Argentina) was about 2 m tall. From Marshall (1978); copyright 1978, 2003 by the artist, Bonnie Dalzell, and used by her kind permission.

Recoveries are, however, still poorly known. Each one has unique aspects, of course: different players, different ecological, climatic and oceanographic settings, and even different continental configurations prevail (e.g. see Hallam and Wignall, 1997; Erwin, 1998, 2001). The challenge is to develop a picture of recovery processes after each major extinction, and to search for common patterns and general rules that emerge despite the differences among the extinction events and post-extinction intervals. As discussed in this section, first steps in this direction have begun to shed some of the easy generalizations or assumptions of earlier work. For example, recoveries are *not* simply relays between dominant groups; they do *not* unfold simultaneously all over

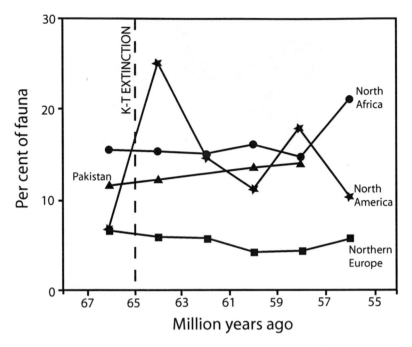

Figure 6.3. A geologically brief evolutionary explosion of bloom taxa occurs immediately after the end-Cretaceous (K–T) mass extinction in North America, but not in northern Europe, north Africa or in the Pakistan/India region. After Jablonski (2003), based on data presented in Jablonski (1998).

the world; and *not* all survivors flourish in the aftermath of mass extinctions.

Geographic variation

Most palaeontological analyses of mass extinctions and their aftermaths have focused either on a very fine scale – a single locality or set of localities – or on global databases such as the monumental compilations by the late Jack Sepkoski (1993, 1997, 2002). Both scales of analysis have been very productive, but the global biota is spatially complex: it is broken up into biogeographic provinces, which are deployed along latitudinal and other gradients, with a biodiversity peak near the equator that falls off towards the poles, and within provinces we see more localized biodiversity hotspots that are becoming a focus for conservation efforts (e.g. see Myers *et al.*, 2000). This spatial structure is also seen in, and may be the result of, large-scale diversity dynamics over geological timescales.

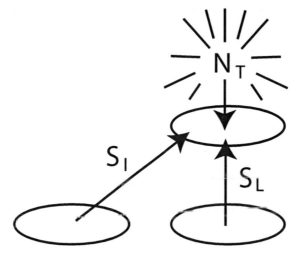

Provincial Fauna = $S_L + S_I + N_T$

Figure 6.4. Within a single region or biogeographic province, the post-extinction fauna (time 2) consists of three components: local survivors (S_L) from time 1, new taxa (N_T), and immigrants (S_I) that had survived elsewhere. The proportions of these components probably vary among regions, extinction events and major groups of organisms.

The recovery from the end-Cretaceous extinction, for example, shows unexpected differences among regions. For example, among North American molluscs, a few groups show a sudden short-lived pulse of diversity not seen elsewhere (Figure 6.3). These 'bloom taxa' (so named because the sudden evolutionary burst and dieback reminded Hansen (1988) of ecological blooms often seen in algal populations in modern lakes after pollution and other stressful events) occur all over the world, but only show this distinct evolutionary pulse along the Gulf Coast of North America.

This is not the only difference that sets the North American recovery apart. This region was also more subject to biological invasions in the aftermath of the mass extinction. After any extinction event, the biota of a given region has three components: local survivors, newly evolved lineages, and invaders that were present elsewhere and enter the region in the wake of the extinction event (Figure 6.4). In the

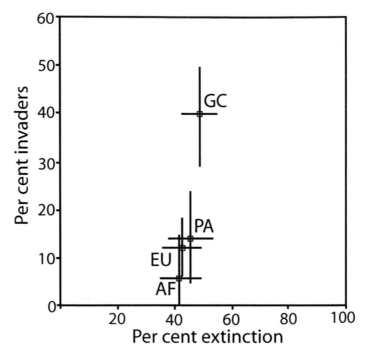

Figure 6.5. Although per cent extinction in marine bivalves shows little variation among regions during the K–T mass extinction, post-extinction fauna in the Gulf Coast of North America contains a significantly higher percentage of invaders, that is, immigrant survivors. Abbreviations: GC, Gulf Coast; PA, Pakistan and India; EU, northern Europe; AF, north Africa; horizontal and vertical bars are 95 per cent statistical confidence limits. From Jablonski (1998).

other post-Cretaceous regions that have been studied in detail, a greater proportion of lineages making up the post-extinction fauna are local survivors, and fewer are post-extinction invaders (Figure 6.5). Why was North America subject to more intense invasions? It is not clear that extinction intensities were any higher here – the dinosaurs went extinct everywhere, and the losses in marine molluscs were about the same here as elsewhere (except perhaps for shallow tropical habitats; see Raup and Jablonski, 1993). Work on present-day invasions suggests that extinction severity and invasion intensity should be linked – the more losses, the more opportunities open to invaders. But Figure 6.4 shows that this expected relationship breaks down at some point, perhaps when the extinction becomes so severe (in this case, 50 per cent of the relatively abundant and widespread genera) that the kinds of victims lost, and not just their number, becomes important. We do not yet know much about this, but it is more than just an academic question. Understanding these patterns in the geological past could be important in light of the accelerating pace of biological invasions owing to human activities (e.g. see Mooney and Hobbs, 2000; Lockwood and McKinney, 2001).

Although body size is not nearly as important in determining victims and survivors of mass extinctions as sometimes thought (Jablonski, 1996), large-bodied molluscs appear to be more likely to spread geographically after the end-Cretaceous mass extinction. This pattern is especially interesting because the same ability of large-bodied forms to shift their geographic range is seen over the past 2 Ma, in response to climatic cycles of the Pleistocene ice ages and interglacials, and in modern seas, where biotic invasions are the result of human activities such as trans-oceanic shipping and mariculture. The consistent outcomes despite such different underlying mechanisms – opportunities opened by mass extinctions, climate changes and human-mediated introductions – suggest that we can discover general rules of biological behaviour in the face of environmental disruption (Jablonski et al., 2003).

Not all survivors are winners

Much of the popular and scientific literature seems to imply that survival alone guarantees evolutionary success in the post-extinction world. However, the fossil record shows that even survivors are subject to a wide range of outcomes (Figure 6.6). Some clades are simply eliminated, as in the end-Cretaceous ammonites, dinosaurs, mosasaurs and pterosaurs. Other clades suffer a set-back in their evolutionary trajectories, but then get back on track. A striking example is seen in the Saunders et al.'s (1999) study on the overall increase in the complexity of sutures in ammonoids through the Palaeozoic (Figure 6.7). The internal walls of the ammonoid shell, which provided chambers like those seen in a Nautilus shell, became increasingly more complicated over Palaeozoic time, except at mass extinctions. It is not clear if such setbacks are a simple by-product of high extinction intensities, or actually represent selection against the traits that were increasing before (and after!) the extinction event.

Still other clades enjoy accelerated diversification following the extinction event. They get a boost from the event, presumably because they were being held in check by incumbents that were pre-empting

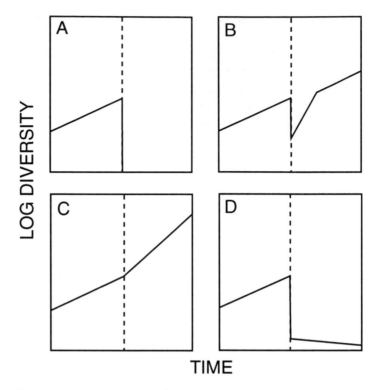

Figure 6.6. Taxonomic or morphological diversity within a clade can follow a wide range of trajectories over time after a mass extinction. After the extinction event indicated by the dashed line, a clade can (A) become extinct, (B) suffer a setback from which it rapidly recovers (see Figure 6.7), (C) accelerate its diversification in response to new opportunities or (D) survive but fail to recover ('Dead Clade Walking'; see Figure 6.8). Diversity is plotted on a logarithmic scale so that exponential diversification appears as a straight line.

ecological opportunities. These are often groups that were minor con-stituents of the pre-extinction biota, as in Cenozoic mammals.

And finally there are clades that survive the extinction but never really get going again. They are marginalized in the aftermath of the extinction, and often dwindle and disappear millions of years after the event. The ones that somehow manage to scrape by until the present day are often called living fossils, like the horseshoe crab, the coelacanth or the pearly nautilus. These clades, the lineages that 'won the extinction but lost the recovery' (Erwin, 1998), can be placed in a category that

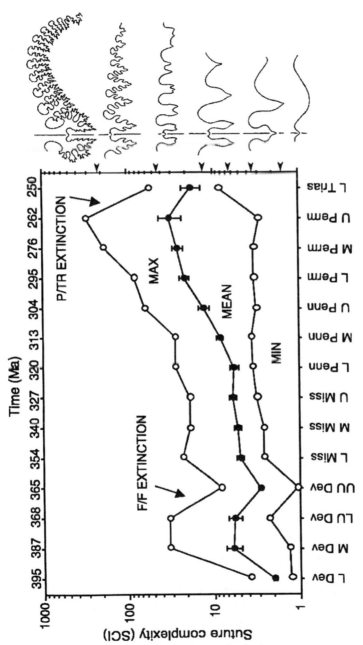

Figure 6.7. The complexity of ammonoid sutures increases through the Palaeozoic Era from the Lower Devonian to the Upper Permian, with setbacks at the Late Devonian (F–F) and end-Permian (P–Tr) mass extinctions. Plot shows mean (bracketed by standard errors), maximum and minimum suture complexity for each time interval, with sutures corresponding to different complexity levels shown at right. From Saunders *et al.* (1999) and used by permission. Ma, million years ago.

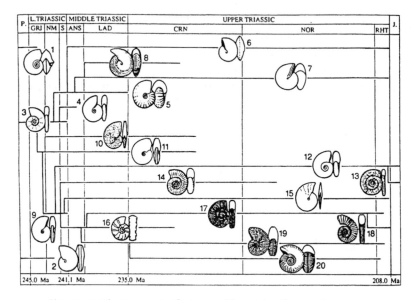

Figure 6.8. Three groups of ammonoids survive the massive
Permo-Triassic extinction (far left, between 'P.' and 'L. Triassic'), but
lineages 1 and 2 fail to recover and so are 'Dead Clades Walking'. Only
lineage 3, the xenodiscids, diversifies during the Triassic, giving rise
to at least 14 major lines with diverse shell forms. From Page (1996) and
used by permission. Ma, million years ago.

I've called 'Dead Clade Walking' (DCW) in homage to an award-winning
film (Jablonski, 2002).

The ammonoids at the end of the Palaeozoic are a good example
of the DCW syndrome. Three groups get across the Palaeozoic–Mesozoic
boundary, but two of them, the otoceratids and the prolecanitids, dwindle out within about 5 Ma. The other survivors, the xenodiscids, went
on to seed the great ammonoid radiations of the Mesozoic (Figure 6.8).

The DCW syndrome appears to play a significant role in post-
extinction dynamics. Suppose we take the narrowest possible definition,
and simply ask, how many lineages that survive a mass extinction event
die out during the next stratigraphic stage, that is within the next
time bin, encompassing the first 5 or 10 Ma of the recovery phase? This
can only underestimate the DCW effect, because many clades might
struggle on for a few million years beyond our post-extinction time
bin and not be picked up by this analysis. Nevertheless, at both the
genus level (the next level up from species) and the much higher and
more inclusive level of taxonomic orders, we see a distinct secondary

concentration of extinctions among survivors in the time interval that immediately follows mass extinction (Figures 6.9, 6.10).

For four of the Big Five mass extinctions, the stages immediately after mass extinctions saw significant attrition among the surviving genera (Figure 6.9). Again, this provides a very conservative estimate of DCW frequencies, because DCW effects clearly do persist for more than one stage, with many bottlenecked taxa nearly static or dwindling for millions of years before disappearing. The one exception to this pattern is the end-Triassic extinction; the lack of sorting after this extinction is consistent with previous observations (Hallam and Wignall, 1997).

The losses of taxonomic orders – major groups like heart urchins, featherstars and stony corals – also tend to be unusually high immediately after mass extinctions, relative to the other non-extinction-stages, but these figures are not as dramatic as for the genera. Figure 6.10 shows the number of orders lost in each of the five mass extinction stages, the number lost in each of the five stages immediately after the mass extinctions (again using that very narrow operational definition of DCW, which will be even more of an underestimate for these larger groups that can dwindle without totally disappearing for many millions of years), and those lost at other times.

The most striking result is simply that the Big Five extinctions account for 35 per cent of order-level extinctions, with a median of 8 orders lost per extinction and a mean of 8.2, very different from the low extinction rates seen for the 70 quieter stages (which have a median extinction intensity of zero and a mean of 1.2 orders). The DCW stages remove another 17 per cent of orders, with a median of 3 and a mean of 5.7 orders lost per DCW stage. The interesting twist here is that all of the DCW losses of orders are concentrated after the three Palaeozoic mass extinctions. That may be because Palaeozoic clades are in general more extinction-prone than post-Palaeozoic ones (Raup and Sepkoski, 1982; Sepkoski, 1998), which would make the short-term sorting of DCWs more detectable at the ordinal level. This is plausible in a general way, but the lack of such extinctions at the ordinal level following the two Mesozoic events despite significant genus-level losses remains an intriguing problem.

As with other aspects of post-Cretaceous recovery, the DCWs are not evenly distributed over the globe. Of the four regions already analysed in Figure 6.3, India/Pakistan contained significantly more DCWs than the others, even though the intensity of post-extinction invasions is unexceptional (Figure 6.11). This is surprising because, as discussed below, failures to recover from mass extinctions are often attributed to

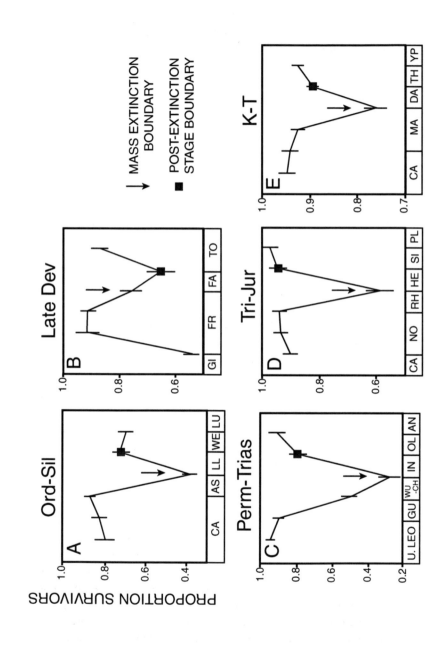

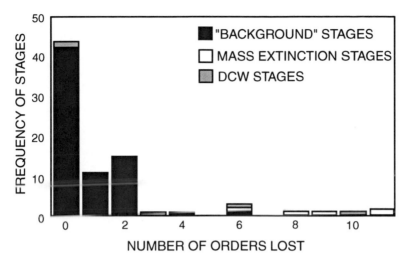

Figure 6.10. More taxonomic orders are lost in the stratigraphic stages
immediately after mass extinction events than in the other stages
showing low, 'background' levels of extinction. Solid bars, number of
orders lost during each of the 'background' stages (45 of the
'background' stages have no order-level extinction); open bars, number
of orders lost in each of the Big Five mass extinctions; hatched bars,
number of orders lost in the stratigraphic stages immediately following
each of the Big Five extinctions. From Jablonski (2002).

biotic interactions like competition or predation. If this were generally
true, then the region with the greatest influx of potential competitors,
predators and parasites should have the most severe post-extinction
losses among native clades. But this pattern is not seen after the end-
Cretaceous extinction. These results suggest that there is a geographic
component to the occurrence of DCWs, so that post-extinction losses

←───

Figure 6.9. Survival of marine animal genera around each of the Big Five
mass extinctions, showing proportion surviving across a stage boundary
only for those genera that had also crossed the preceding stage
boundary. In the Ordovician-Silurian (A), Late Devonian (B),
Permo-Trassic (C) and end-Cretaceous (K–T; E) extinctions, the *genera* that
survived the extinction itself (shown as an arrow) had significantly lower
survivorship in the immediately following stage or stages (shown as a
square) than was seen for taxa at pre-extinction stage boundaries.
Vertical bars are 95 per cent statistical confidence limits; two-letter
codes along the bottom of each plot are abbreviations of the names for
those stratigraphic stages. From Jablonski (2002).

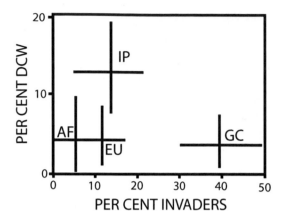

Figure 6.11. The occurrence of 'Dead Clades Walking' (DCW) (genera of marine clams and snails) in each region is unrelated to the per cent invaders in each post-extinction biota after the K–T mass extinction. Abbreviations: AF, north Africa; EU, northern Europe; GC, Gulf Coast; IP, India and Pakistan; horizontal and vertical bars are 95 per cent statistical confidence limits. From Jablonski (2002).

might reflect survivorship influenced by where the organisms lived, and not just the biological adaptations of the biota.

There are many possible mechanisms for the DCW syndrome. First, individual examples have to be checked carefully to rule out preservational artefacts, such as reworked fossils from older, pre-extinction deposits. Fortunately, such reworking is usually recognizable for large fossils like clamshells or dinosaur bones, although it can be much more of a problem for microfossils like pollen grains or foraminifera. More generally, the concentration of DCWs in the intervals immediately after extinctions is unlikely to be a large-scale collecting bias either. For a host of reasons, the number of sites yielding fossils generally goes down after a mass extinction (Foote, 2003; Chapter 3), and as Twitchett (2001) discussed, this decrease in sampling intensity is accompanied by a 'fossilization low' typified by the absence of lineages known to have survived the mass extinction because they turn up in the fossil record again sometime after the extinction event (this is the 'Lazarus effect', see Jablonski, 1986; Chapter 1). If anything, these sampling biases immediately after mass extinctions should artificially *reduce* the number of DCWs, because their ranges will be truncated by sampling failure so that they will appear to die out during the extinction itself (see Foote, 2000).

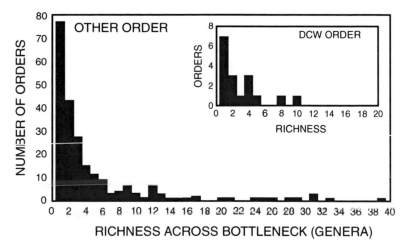

Figure 6.12. The bottlenecks suffered by the 'Dead Clade Walking' (DCW) orders of marine invertebrates after the Big Five mass extinctions, as measured by the number of genera within each order that survives the mass extinction, are no more severe than seen for the other orders. Compare the bottleneck sizes for the DCW orders against all the other surviving orders. From Jablonski (2002).

A second explanation for the DCW syndrome is a bottleneck effect, the almost inevitable aftermath of a great reduction in the number of species within a given clade. Once the clade is knocked down to a very small number of species, putting it through a severe bottleneck, it is just a matter of time before bad luck or some relatively mild disturbance takes out the last survivors and the clade is gone for good. This is perfectly plausible in principle, but it does seem to apply here, judging by two complementary lines of evidence. First, the DCW orders were no more severely bottlenecked by the extinction than other clades that went on to diversify in the recovery interval (Figure 6.12). Second, there is no correlation between the size of the bottleneck and how long the order persists after the extinction (Figure 6.13), contrary to what we would expect from bottleneck effects. These results suggest that the loss of DCWs must be explained by something other than the operation of chance and the simple laws of probability. They didn't fall, they were pushed.

That push could have come from later environmental disturbances. For example, the concentration of DCWs in India/Pakistan might very well be related to the destruction of the north Indian continental shelf as India ploughed into Asia to produce the Himalayan

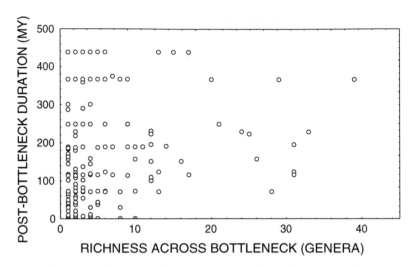

Figure 6.13. The survival of an order in millions of years (Ma) after one of the Big Five mass extinctions is unrelated to the size of the bottleneck that it went through during the mass extinction. If bottleneck effects were important we should see a positive relationship between the size of an extinction bottleneck, with orders that went through the smallest bottlenecks being the most extinction-prone. From Jablonski (2002).

Mountains and the Tibetan Plateau. The collision occurred about 10 Ma after the end-Cretaceous extinction, so the timing is about right. Further, if we repeat the analysis without the genera that are restricted to the India and Pakistan region (since they are the ones most vulnerable to the effects of the collision), the DCW proportion drops down to about 6 per cent, and is thus the same as all the other regions.

Harsh environmental conditions that persist after the extinction events themselves may be responsible for other losses in the aftermath of mass extinctions, and even for delayed recoveries. For example, Twitchett (2001) suggested that the end-Permian mass extinction was followed by an extended interval of lower primary productivity in the oceans, with the phytoplankton that formed the base of the food chain remaining in a state of collapse. More work is needed to test this intriguing idea against the many alternatives, from low-oxygen conditions to delays in re-assembling ecological communities after such massive losses of biodiversity (see Erwin *et al.*, 2002). A better understanding of the environmental challenges faced by the lineages in the post-extinction world will be crucial to evaluating the DCW phenomenon.

One last explanation for the DCW syndrome, the one most often invoked and the most difficult to test rigorously, involves interactions with other surviving groups. After all, there are bound to be winners and losers in the scramble for resources in the open ecological opportunities after extinction events. Perhaps the failure of the prolecanitid ammonoids to rediversify, for example, resulted from competition with the xenodiscids. But such a process cannot be demonstrated simply by tracking the waxing and waning of major groups. For example, the decline of the dinosaurs and the diversification of the mammals was once viewed in terms of the competitive superiority of the mammals. With the discovery that dinosaurs disappeared at virtually the same time as many land plants, marine invertebrates and oceanic phytoplankton, and that this upheaval coincided with the formation of the Chicxulub crater, the competition hypothesis has been discarded in favour of the hypothesis that the replacement was driven by an external force, probably one or more asteroid impacts, and not ecological interactions.

Testing ecological explanations requires more than simply rejecting the non-biological alternatives just mentioned. Additional information is needed, such as the life habits, biogeography, and relative abundances of the DCWs and the rivals and other enemies hypothesized to be damping or reversing their diversification. Whatever the conclusion on driving mechanisms, and work is just beginning here, it is clear the recovery phase may sometimes be as important as the extinction itself in shaping the new biota, and in setting new evolutionary directions. Not all survivors are winners.

SUMMARY AND IMPLICATIONS FOR THE FUTURE

If we take a sufficiently long view – millions of years instead of the human timescales of years and decades – then we can recognize that extinction is a double-edged sword (Jablonski, 2001). By definition it removes clades, adaptations, and genetic variation; impoverishment is the legacy of extinction on human timescales. But over geological timescales it can reshape the evolutionary landscape in more creative ways, thanks to the evolutionary opportunities opened up by the demise of dominant groups and the post-extinction sorting of survivors.

Extinction never completely re-sets the evolutionary clock. Some groups sail right through major extinction events, or get through with set-backs that are soon recouped. This imparts considerable biological continuity across even the biggest extinction events. At least two-thirds of mammalian history had passed before they got their great opportunity to diversify. And while the level of evolutionary creativity displayed by mammals when they got their chance is impressive – bats, whales, and eventually giraffes, anteaters and humans – none of these forms were as radically new as the marine invertebrate body plans that diversified around the very start of the Palaeozoic Era during the Cambrian explosion. The world has seen only one evolutionary event on the Cambrian scale (see Erwin *et al.*, 1987; Valentine *et al.*, 1999; Carroll and Knoll, 1999; Valentine, 2002; Jablonski, 2003). Several hypotheses are actively being pursued to explain the failure of recoveries to pro-duce the kinds of novel body plans seen during the Cambrian explosion. Although more work is needed to test the idea that the embryology and genetics of Cambrian organisms were more flexible than in later times, the strongest support appears to be for the view that the survivors of each extinction event, although jostling for position and making evo-lutionary shifts in response to new opportunities, were operating in a world that was never really emptied out, ecologically speaking. In other words, incumbency effects damped post-extinction diversification, com-pared to the Cambrian explosion (see Erwin, 2001).

What are the implications of these insights into past extinctions and recoveries for the future of life on Earth? Although work is just beginning in this field, at least five lessons are becoming apparent (see Jablonski, 1995, 2001; Sepkoski, 1997; Erwin, 2001; Novacek and Cleland, 2001).

First, mass extinctions do happen. Clades and communities are not infinitely resilient but can be pushed to the breaking point. Direct comparisons of past and present-day extinctions are difficult: the causes are very different, of course, as are the spatial and temporal scales of the extinctions (Jablonski, 1995, 2001). But we can turn to the fossil record for lessons on thresholds, on the relative sensitivity of different groups at a single time, and on the qualitative – not just the quantitative – differences in the effects of extinction events of different magnitude.

Second, survival during mass extinctions may not be strongly tied to biological success during quieter times. This is probably the reason that well-established incumbents are lost, and not just the more vul-nerable, marginal players in the evolutionary game as it stands dur-ing times of low extinction intensities. We need to learn much more about this apparent shift in the rules of survival: which ecological types, clades, geographic regions, or other groupings are at greatest risk of a switch in survivorship patterns, and is the trigger for that switch con-stant among groups or time intervals?

Third, because mass extinctions are especially severe among local-ized, endemic clades, they tend to homogenize the world's fauna and flora. This removal of endemics and specialists creates a world more heavily dominated by the ubiquitous weedy species that reproduce rapidly and are superb invaders of new regions: the rats and the cockroaches tend to survive and spread, not the gazelles and the orchids – or the species most helpful to humans such as potential medicinal or food plants. Because widespread species tend to produce new descendant species at low rates (Jablonski and Roy, 2003), the increasing domination of the world biota by such clades would probably result in a slowing of speciation rates – that is, a reduction in the pace of species production, even as extinction rates are accelerating.

Fourth, the evolutionary recovery is slow on human timescales. It might seem from this chapter and others that catastrophe and renewal go hand-in-hand, so that there really is nothing to worry about because Mother Nature will right herself eventually. That may be true over the broad sweep of geological time, but a key message of the fossil record is the great disparity between the time scales of extinction and recovery. Extinctions can be fast or slow, but it always takes time to rebuild ecosystems and to exploit vacant niches. The mammals did inherit the Earth, but it took them over 10 Ma to fan out into the niches vacated by the victims of the end-Cretaceous extinction – a bolt of evolutionary lightning for a geologist, but unimaginably slow compared to a human lifetime and a long time compared to the total lifespan of our species so far.

Finally, recoveries are not only slow but unpredictable. Not all survivors are winners, but we do not know much about that piece of the puzzle. If the long-term goal of conservation biology is to save not only individual species or ecosystems, but to allow the continuity of the evolutionary process in today's beleaguered biosphere, then we need to understand which clades are vulnerable to failure during the recovery phase. This presumes, of course, that our species actually permits a recovery phase (see Myers and Knoll, 2001).

In order to understand the dynamics of biodiversity, the rise and fall of the great evolutionary dynasties on geological timescales – for example, why this chapter is being written by a furry mammal instead of a scaly dinosaur or a feathery bird – we need to understand extinctions and their complex aftermath. Because they remove incumbents, overturn the rules of survival that prevail in quieter times, and unleash a scramble for post-extinction opportunities that can produce bursts of evolutionary novelty, mass extinctions have played pivotal roles in the history of life. The interplay between these destructive and creative aspects of mass extinction is only beginning to be explored. We are, however, getting a better sense of how these processes work, and seeing that the fates of lineages do not always fall into the simple categories of woeful victim or triumphant (and inevitable) survivor. But even here our picture is still very incomplete. We need to know whether recovery is proportional to the severity of the extinction event – do lesser extinction pulses that fall short of the Big Five also remove at least some dominants and change the rules for some groups, habitats or regions, or is there a threshold that must be crossed before the dynamics discussed here come into play? This seems a promising research direction. The fossil record is rich in evolutionary experiments, and clearly can be a source of much insight, not only into the workings of past life, but into the potential fates of species living today.

ACKNOWLEDGMENTS

I thank J. William Schopf and Paul D. Taylor for inviting me to participate in the stimulating symposium where this paper was read, and am grateful for their editorial patience and acumen. Susan Kidwell provided, as always, an extremely valuable review. Thanks also to Bonnie Dalzell for permission to use her phorusrhacid illustration. The research reported here was supported by the National Science Foundation, and much of the work was made possible by my appointment as an Honorary Research Fellow at The Natural History Museum, London.

REFERENCES

Balinski, A., Olempska, E. and G. Racki, G. (Eds.), 2002. Biotic responses to the Late Devonian global events. *Acta Palaeontologica Polonica* **47**: 186–404.

Bambach, R. K., Knoll, A. H. and Sepkoski, J. J. Jr, 2002. Anatomical and ecological constraints on Phanerozoic animal diversity in the marine realm. *Proceedings of the National Academy of Sciences, USA* **99**: 6854–6859.

Benton, M. J., 1996. On the nonprevalence of competitive replacement in the evolution of tetrapods. In: D. Jablonski, D. H. Erwin, and J. H. Lipps, (Eds.), *Evolutionary Paleobiology*. Chicago: University of Chicago Press, pp. 185–210.

Carroll, S. B. and Knoll, A. H., 1999. Early animal evolution: emerging views from comparative biology and geology. *Science* **284**: 2129–2137.

Copper, P., 2002. Reef development at the Frasnian/Famennian mass extinction boundary. *Palaeogeography, Palaeoclimatology, Palaeoecology* **181**: 27–65.

Dyke, G. J., 2001. The evolutionary radiation of modern birds: systematics and patterns of diversification. *Geological Journal* **36**: 305–315.

Erwin, D. H., 1989. Regional paleoecology of Permian gastropod genera, southwestern United States and the end-Permian mass extinction. *Palaios* **4**: 424–438.

1998. The end and the beginning: recoveries from mass extinctions. *Trends in Ecology and Evolution* **13**: 344–349.

2001. Lessons from the past: Biotic recoveries from mass extinctions. *Proceedings of the National Academy of Sciences, USA* **98**: 5399–5403.

Erwin, D. H., Bowring, S. A., and Yugan, J., 2002. End-Permian mass extinctions: a review. *Geological Society of America Special Paper* **356**: 363–383.

Erwin, D. H., Valentine, J. W., and Sepkoski, J. J. Jr, 1987. A comparative study of diversification events: the early Paleozoic versus the Mesozoic. *Evolution* **41**: 1177–1186.

Feduccia, A., 1999. *The Origin and Evolution of Birds*, 2nd edn. New Haven. Yale University Press.

2003. 'Big bang' for Tertiary birds? *Trends in Ecology and Evolution* **18**: 172–176.

Foote, M., 2000. Origination and extinction components of taxonomic diversity: general problems. *Paleobiology* **26** (Suppl. to No. 4): 74–102.

2003. Origination and extinction through the Phanerozoic: a new approach. *Journal of Geology* **111**: 125–148.

Gould, S. J., 2002. *The Structure of Evolutionary Theory*. Cambridge, MA: Harvard University Press.

Hallam, A. and Wignall, P. B., 1997. *Mass Extinctions and their Aftermath*. Oxford: Oxford University Press.

Hansen, T. A., 1988. Early Tertiary radiation of marine mollusks and the long-term effects of the Cretaceous–Tertiary extinction. *Paleobiology* **14**: 37–51.

Hansen, T. A., Upshaw, B., Kauffman, E. G. and Gose, W., 1993. Patterns of molluscan extinction and recovery across the Cretaceous–Tertiary boundary in east Texas: report on new outcrops. *Cretaceous Research* **14**: 685–706.

Harper, D. A. T. and Rong, J.-Y., 2001. Palaeozoic brachiopod extinctions, survival and recovery: patterns within the rhynchonelliformeans. *Geological Journal* **36**: 317–328.

Harries, P. J., 1999. Repopulations from Cretaceous mass extinctions: environmental and/or evolutionary controls? *Geological Society of America Special Paper* **332**: 345–364.

Jablonski, D., 1986. Causes and consequences of mass extinctions: a comparative approach. In: D. K. Elliott (Ed.), *Dynamics of Extinction*. New York: Wiley, pp. 183–229.

1989. The biology of mass extinction: a paleontological view. *Philosophical Transactions of the Royal Society of London* **B325**: 357–368.

1995. Extinction in the fossil record. In: R. M. May and J. H. Lawton (Eds.), *Extinction Rates*. Oxford: Oxford University Press, pp. 25–44.

1996. Body size and macroevolution. In: D. Jablonski, D. H. Erwin and J. H. Lipps (Eds.), *Evolutionary Paleobiology*. Chicago: University of Chicago Press, pp. 256–289.

1998. Geographic variation in the molluscan recovery from the end-Cretaceous extinction. *Science* **279**: 1327–1330.

2001. Lessons from the past: evolutionary impacts of mass extinctions. *Proceedings of the National Academy of Sciences, USA* **98**: 5393–5398.

2002. Dead clade walking: survival without recovery after mass extinctions. *Proceedings of the National Academy of Sciences, USA* **99**: 8139–8144.

2003. The interplay of physical and biotic factors in macroevolution. In: L. Rothschild and A. Lister (Eds.), *Evolution on Planet Earth*. London: Academic Press, pp. 235–252.

Jablonski, D. and Raup, D. M., 1995. Selectivity of end-Cretaceous marine bivalve extinctions. *Science* **268**: 389–391.

Jablonski, D. and Roy, K., 2003. Geographical range and speciation in fossil and living mollusks. *Proceedings of the Royal Society of London* **B270**: 401–406.

Jablonski, D., Roy, K. and Valentine, J. W., 2003. Evolutionary macroecology and the fossil record. In: T. M. Blackburn and K. J. Gaston (Eds.), *Macroecology: Concepts and Consequences*. Oxford: Blackwell Science, pp. 368–390.

Kitchell, J. A., Clark, D. L. and Gombos, A. M., 1986. Biological selectivity of extinction: a link between background and mass extinction. *Palaios* **1**: 504–511.

Lockwood, R., 2003. Abundance not linked to survival across the end-Cretaceous mass extinction: patterns in North American bivalves. *Proceedings of the National Academy of Sciences, USA* **100**: 2478–2482.

Lockwood, J. L. and McKinney, M. L. (Eds.), 2001. *Biotic Homogenization*. New York: Kluwer/Plenum.

Marshall, L. G., 1978. The terror bird. *Field Museum of Natural History Bulletin* **49** (9): 6–15.

McGhee, G. R., Jr, 1996. *The Late Devonian Mass Extinction*. New York: Columbia University Press.

Mooney, H. A. and Hobbs, R. J. (Eds.), 2000. *Invasive Species in a Changing World*. Washington DC: Island Press.

Myers, N. and Knoll A. H., 2001. The biotic crisis and the future of evolution. *Proceedings of the National Academy of Sciences, USA* **98**: 5389–5392.

Myers, N., Mittermeier, R. A., Mittermeier, C. G., da Fonseca, G. A. B. and Kent, J., 2000. Biodiversity hotspots for conservation priorities. *Nature* **403**: 853–858.

Norris, R. D., 1991. Biased extinction and evolutionary trends. *Paleobiology* **17**: 388–399.

Norris, R. D., 1992. Extinction selectivity and ecology in planktonic foraminifera. *Palaeogeography, Palaeoclimatology, Palaeoecology* **95**: 1–17.

Novacek, M. and Cleland, E. E., 2001. The current biodiversity extinction event: scenarios for mitigation and recovery. *Proceedings of the National Academy of Sciences, USA* **98**: 5466–5470.

Page, K. N., 1996. Mesozoic ammonoids in space and time. In: N. H. Landman, K. Tanabe, and R. A. Davis (Eds.), *Ammonoid Paleobiology*. New York: Plenum Press, pp. 755–794.

Plotnick, R. E. and Baumiller, T. K., 2000. Invention by evolution: functional analysis in paleobiology. *Paleobiology* 26 (Suppl. to No. 4): 305–323.

Racki, G., 1999. Silica-secreting biota and mass extinctions: survival patterns and processes. *Palaeogeography, Palaeoclimatology, Palaeoecology* 154: 107–132.

Racki, G. and House, M. R. (Eds.), 2002. Late Devonian biotic crisis: ecological, depositional and geochemical records. *Palaeogeography, Palaeoclimatology, Palaeoecology* 181: 1–374.

Raup, D. M., 1991. A kill curve for Phanerozoic marine species. *Paleobiology* 17: 37–48.

1994. The role of extinction in evolution. *Proceedings of the National Academy of Sciences, USA* 91: 6758–6763.

Raup, D. M. and Jablonski, D., 1993. Geography of end-Cretaceous marine bivalve extinctions. *Science* 260: 971–973.

Raup, D. M. and Sepkoski, J. J., Jr, 1982. Mass extinctions in the marine fossil record. *Science* 215: 1501–1503.

Rosen, B. R., 2000. Algal symbiosis, and the collapse and recovery of reef communities: Lazarus corals across the K–T boundary. In: S. J. Culver and P. F. Rawson (Eds.), *Biotic Responses to Global Change*. Cambridge: Cambridge University Press, pp. 164–180.

Saunders, W. B., Work, D. M. and Nikolaeva, S. V., 1999. Evolution of complexity in Paleozoic ammonoid sutures. *Science* 286: 760–763.

Sepkoski, J. J., Jr, 1993. Ten years in the library: new data confirm paleontological patterns. *Paleobiology* 19: 43–51.

1997. Biodiversity: past, present, and future. *Journal of Paleontology* 71: 533–539.

1998. Rates of speciation in the fossil record. *Philosophical Transactions of the Royal Society of London* B353: 315–326.

2002. A compendium of fossil marine animal genera. *Bulletins of American Paleontology* 363: 1–560.

Smith, A. B. and Jeffrey, C. H., 1998. Selectivity of extinction among sea urchins at the end of the Cretaceous Period. *Nature* 392: 69–71.

Twitchett, R. J., 2001. Incompleteness of the Permian–Triassic fossil record: a consequence of productivity decline? *Geological Journal* 36: 341–353.

Valentine, J. W., 2002. Prelude to the Cambrian explosion. *Annual Review of Earth and Planetary Sciences* 30: 285–306.

Valentine, J. W., Jablonski, D. and Erwin, D. H., 1999. Fossils, molecules and embryos: new perspectives on the Cambrian explosion. *Development* 126: 851–859.

Witmer, L. M. and Rose, K. D., 1991. Biomechanics of the jaw apparatus of the gigantic Eocene bird *Diatryma*: implications for diet and mode of life. *Paleobiology* 17: 95–120.

Glossary

Accretionary prisms Wedges of deformed sediment, often several hundred kilometres in lateral extent, formed from material scraped off the oceanic crust as it descends into the mantle.

Acritarch An artificial taxonomic group that includes Precambrian and Phanerozoic organic-walled, commonly spheroidal, algal-like single-celled fossils of uncertain biological affinities.

Adaptive radiation The evolution of a species into a group of species adapted to different niches.

Age of Evident Life Informal name for the Phanerozoic Eon.

Age of Microscopic Life Informal name for the Precambrian Eon.

Alga Any of diverse types of eukaryotic photoautotrophic single-celled protists, such as phytoplankton, or many-celled seaweeds.

Ammonoids (including ammonites) Very common extinct group of marine shellfish, related to squid and octopus, that had whorled and chambered shells.

Angiosperm Any member of the taxonomic group (Angiospermae) that consists of flowering plants.

Anoxic waters Waters lacking oxygen, which are lethal to all non-microbial life.

Archaea Any of diverse microbes of the Archaeal domain.

Archaeal domain Together with Bacteria and Eucarya, one of three superkingdom-like primary branches of the Tree of Life.

Archean Era The older era of the Precambrian Eon, extending from Earth's formation 4500 Ma ago to the beginning of the Proterozoic 2500 Ma ago; together, the Archean and Proterozoic Eras comprise the Precambrian Eon.

Bacterial domain Together with Archaea and Eucarya, one of three superkingdom-like primary branches of the Tree of Life.

Bacterium Any of diverse prokaryotes, including cyanobacteria, of the bacterial domain.

Banded iron formation Chemically deposited cherty sedimentary rock, usually thinly bedded and containing more than 15 per cent iron.

Big Five   Informal term commonly used to describe the five largest extinction events of the fossil record, the end-Ordovician, Late Devonian (Frasnian–Famennian), end-Permian, end Triassic and end-Cretaceous mass extinctions.

Bloom taxa   Taxa exhibiting a sudden evolutionary burst followed by a dieback.

Cambrian Period   The earliest geological period of the Phanerozoic Eon of Earth's history, extending from 543 to 495 Ma ago.

Catastrophism   The doctrine, expounded by Georges Cuvier, that catastrophic processes are of great importance in shaping the geological record (cf. Uniformitarianism).

Cenomanian–Turonian   Stages of the Cretaceous, the boundary between which at around 90 Ma ago marks a minor extinction event.

Cenozoic Era   Youngest of three eras of the Phanerozoic Eon of Earth's history, extending from the end of the Mesozoic Era, 65 Ma ago, to the present.

Chert gap   A characteristic of the marine sedimentary rock record of the Early Triassic, where rocks made of the skeletons of radiolarians and siliceous sponges (chert) are rare to absent.

Chicxulub   Small town on the northern coast of the Yucatan Peninsula, Mexico, close to the centre of a meteorite impact site formed 65 Ma ago, at the end of the Cretaceous. The crater is now entirely infilled by sediment, but is thought to have been originally at least 180 km in diameter.

Chromosome   Elongate structures, in eukaryotes occurring in the cell nucleus, that contain the hereditary molecule, DNA.

Chroococcaceae   A taxonomic family of simple, unicellular or colonial, spheroidal cyanobacteria.

Clade   A natural, monophyletic group of organisms including an ancestral species plus all of its descendant species.

Coal gap   A characteristic of the terrestrial sedimentary rock record of the Early Triassic, where rocks made of the remains of plants that lived in wetland ecosystems (coal) are rare to absent.

Coccolithophores   A group of phytoplankton that constructs calcite spheres, called coccoliths, out of tiny plates.

Continental flood basalt province   A large igneous province on a continent formed by a mantle plume.

Cyanobacterium   Any of a diverse group (Cyanobacteria) of prokaryotic microorganisms capable of oxygen-producing photosynthesis (the group in older classifications are termed blue-green algae).

Dead Clade Walking (DCW)   Clade that survives an extinction event but fails to rediversify.

Deccan Traps   Vast province of basaltic lava flows covering a large area of northwest India. They were erupted in a short interval bracketing the K–T boundary. Their original volume may have approached 4×10^6 km^3.

Deoxyribonucleic acid (DNA)   The genetic information-containing molecule of cells, a double-stranded nucleic acid made up of nucleotides that contain a nitrogenous base (adenine, guanine, thymine, or cytosine), deoxyribose sugar and a phosphate group.

Diploid   A cell or organism possessing two sets of chromosomes such that every gene is present as two copies.

Domain   A superkingdom-like primary branch of the Tree of Life.

Ecologic generalist   An organism capable of living in ecologically diverse habitats, e.g. many kinds of cyanobacteria.

Ecologic specialist   An organism well adapted to an ecologically limited habitat, such as most eukaryotes.

Ecology   The science that deals with the interrelations among organisms inhabiting a common environment and between these organisms and the environment.

Endolith   Any of diverse organisms (commonly prokaryotic, algal, or fungal) that live within a rock or consolidated soil crust.

Entophysalidaceae   A taxonomic family of predominantly colonial, mucilage-enclosed ellipsoidal cyanobacteria.

Eucaryal domain   Together with Bacteria and Archaea, one of three superkingdom-like primary branches of the Tree of Life.

Eukaryote   Any of a taxonomic group (the Eucarya) of organisms composed of one or more nucleus-containing cells; any member of the Eucaryal domain such as a protist, fungus, plant or animal.

Evolutionary stasis   Lack of evolutionary change over geologically long periods.

Extinction selectivity   Different patterns of extinction and survival among evolutionary clades or according to biological characteristics of the organisms.

Extraterrestrial bolide impact   The impact on the Earth's surface of an extraterrestrial body such as an asteroid or comet.

Extremeophile   Microbes such as many archaeans that tolerate exceptionally high-temperature acidic environments.

Facultative aerobe   Any of various prokaryotes, usually bacterial, capable of aerobic respiration but that can also grow in the absence of molecular oxygen (O_2) by anaerobic metabolism.

Filament   In microbiology, a collective term for the cylindrical external sheath and cellular internal trichome of a filamentous prokaryote.

Flood basalt   Low viscosity lava erupted from fissures in the Earth's surface forming sheet flows and resulting in tracts of the igneous rock basalt that cover tens of thousands of square kilometres.

Foraminifera   Important group of marine protists that secretes tiny chambered skeletons of calcite; includes forms that are planktonic, although the majority live on the seafloor.

Frasnian–Famennian.   The last two stratigraphical stages of the Devonian whose boundary, at around 365 Ma ago, marks one of the Big Five mass extinctions.

Fullerenes   Large organic molecules consisting of spheres constructed of 60 carbon atoms. They are thought to occur in comets but they can also be produce by lightning strikes in soil.

Gamete   Haploid female (egg) or male (sperm) sex cell, in animals formed by meiosis and in plants by mitotic division of haploid cells derived from meiotically produced spores.

Gas hydrate   Ice formed in cold conditions, under considerable pressure, is able to trap alkane gases, principally methane, and thus form gas hydrates. At low latitudes they are found within continental slope sediments and at higher latitudes they occur in shallower waters.

Gene   A segment of DNA containing information for production of a protein or RNA (ribonucleic acid) molecule.

Greenhouse gas   An atmospheric gas that causes the atmosphere to retain greater amounts of heat as its concentration in the atmosphere increases.

Guadalupian   Name commonly used by North American geologists to describe an interval of the Middle Permian, roughly 265 Ma ago (see also Maokouan).

Gymnosperm   Any member of the taxonomic group (Gymnospermae) that consists of plants having naked seeds, such as the conifers, cycads and *Ginkgo*.

Haploid   In sexually reproducing organisms, the chromosomal complement present in sperm or egg.

Hyellaceae   A taxonomic family of predominantly endolithic cyanobacteria.

Hypsometry   Term concerned with the distribution of continental elevation, for example a continent lacking large mountains, such as Australia, is said to have a low hypsometric gradient.

Incumbents   Groups that are well-established in their ecological roles and thus pre-empt others from those lifestyles.

Invertebrate   Any of the diverse animals that lack backbones.

K–T mass extinction   The major extinction that occurred at the end of the Cretaceous period about 65 Ma ago. Cretaceous is abbreviated to K rather than C to avoid confusion with a C–T (Cenomanian–Turonian) boundary earlier in the Cretaceous. The letter K stands for Cretaceous in the Greek form *Kreta* and the German form *Kreide*, and T for Tertiary, the subera of geological time succeeding the Cretaceous. As Tertiary is no longer recognized as a formal stratigraphical term, the extinction is now sometimes referred to as the KP mass extinction, P signifying Paleocene.

Large igneous provinces   Enormous regions on the Earth's surface formed by prodigious outpourings of flood basalts, such as the Siberian Traps.

Lazarus Effect/Taxon   Temporary disappearance from the stratigraphical record followed by reappearance in younger rocks; a result of the incompleteness of the fossil record.

Living fossil   A living organism (for example, a horseshoe crab, a *Ginkgo* tree, or any of various cyanobacteria) that has remained essentially unchanged in morphology for a long interval of geological time.

Ma   Mega anna, one million (1 × 10^6) years.

Mantle plume   A column of melted igneous rock which rises from the mantle through the crust to the Earth's surface, and which is the cause of flood basalts and large igneous provinces.

Maokouan   Name commonly used to describe an interval of the Middle Permian, roughly 265 Ma ago (see also Guadalupian).

Mass extinction   Geologically brief interval when extinction was significantly above the background level, and which typically had a catastrophic effect broadly across the Earth's biota.

Meiosis   The process of nuclear division that reduces the number of chromosomes from the diploid to the haploid number in each of four product cells (in animals, sperm or egg).

Mesozoic Era   The middle of the three major Phanerozoic time intervals, ranging from the beginning of the Triassic (251 Ma ago) to the end of the Cretaceous (65 Ma ago).

Microbe   Informal term for any of diverse types of prokaryotic bacteria or archaeans.

Micrometre (μm)   A unit of length, one-millionth (10^{-6}) of a metre (or one-thousandth of a millimetre).

Mitosis   A type of division of the cell nucleus resulting in formation of two daughter cells, each a genetic copy (clone) of the parent cell; in unicellular eukaryotes, a type of nonsexual reproduction.

Mollusc   Any of an animal phylum (Mollusca) characterized by a large muscular foot and a mantle that normally secretes a shell or less commonly spicules, such as a snail, clam or squid.

Mutation   Any change in the sequence of nucleotides (adenine-, guanine-, thymine- or cytosine-containing chemical structures) of a gene.

Nonsexual   With reference to organisms that lack capability to reproduce sexually.

Nucleus   In eukaryotes, a membrane-enclosed organelle that contains the chromosomes.

Oceanic anoxic events   Intervals of geological time marked by the widespread development of anoxic waters in the oceans; often associated with mass extinction events.

Oscillatoriaceae   A taxonomic family of simple filamentous cyanobacteria that lacks heterocysts, a particular kind of specialized cell.

Ozone   A triatomic form of oxygen, O$_3$, formed naturally in the upper atmosphere.

Palaeozoic Era   The oldest of three geological eras of the Phanerozoic Eon of Earth history, extending from the end of the Proterozoic Era of the Precambrian Eon, 543 Ma ago, to the beginning of the Mesozoic Era, 251 Ma ago.

Pangea   The supercontinent assembled by plate tectonic processes during the middle part the of Phanerozoic Eon.

Panthalassa   The world ocean during the time in Earth's history when Pangea existed as a single supercontinent.

Paraphyletic   An artificial group of taxa comprising an ancestral species and some, but not all, of its descendants.

Period (geological)   A formal division of geological time longer than an epoch and included in an era.

Phanerozoic Eon   The younger of two principal divisions (eons) of Earth's history, extending from the beginning of the Cambrian (about 543 Ma ago) to the present day; most fossils of multicellular organisms come from the Phanerozoic.

Photic zone   The surface layers of oceans or lakes where sufficient light penetrates to support photosynthesis.

Photosynthesis   The metabolic process carried out by photosynthetic bacteria, cyanobacteria, algae and plants in which light energy is converted to chemical energy and stored in molecules of biosynthesized carbohydrates.

Photosynthetic bacterium   Any of diverse types of bacteria capable of anoxygenic photosynthesis.

Photosynthetic space   The surface area or volume within the photic zone where photosynthesis can occur.

Phytoplankton   Plant-like plankton, such as single-celled algae, that float in the upper layers of the open ocean.

Plankton   Organisms inhabiting the surface layers of a sea or lake, such as small drifting algae, protozoans and animals.

Planktonic foraminifera   Shelled protozoans that float in the open ocean.

Pleurocapsaceae   A family of predominantly colonial, mucilage-enclosed ellipsoidal cyanobacteria.

Precambrian Eon   The older of two principal divisions (eons) of Earth's history, extending from the formation of the planet, 4500 Ma ago, to the beginning of the Cambrian Period, 543 Ma ago; the Precambrian and the younger Phanerozoic Eon comprise all geological time.

Prokaryote   Any of diverse types of non-nucleated microorganisms of the Archaea and Bacteria.

Proterozoic Era   The younger era of the Precambrian Eon, extending from the end of the Archean Era, 2500 Ma ago, to the beginning of the earliest (Cambrian) period of the Phanerozoic Eon, 543 Ma ago; together, the Proterozoic and the Archean Eras comprise the Precambrian Eon.

Protist   A general term for single-celled plant- or animal-like eukaryotes.

Protozoans   Animal-like protists.

Pseudoextinction   False extinction resulting from taxonomic procedure. For example, when one species in a lineage evolves into another, a change of name occurs but there is no termination of a branch in the evolutionary tree.

Radiolarians   Marine plankton, with a long history stretching from the Cambrian to the present day, that construct delicate skeletons made of a lattice work of silica. They are generally less than a millimetre in size.

Reef eclipse A time in Earth history when reefs built by colonial organisms, such as corals, are absent to rare over millions of years.

Reefs Seafloor structures built by the mineralized skeletons of a variety of marine organisms, commonly corals, ranging in areal extent from small patch reefs to huge barrier systems that fringe coastlines for many kilometres.

Regression Seaward movement of the coastline. This often happens during sea-level falls.

Seed plant Any of diverse 'higher land plants' such as gymnosperms and angiosperms that produce seeds.

Sexual reproduction In eukaryotes, a process of reproduction involving the formation of gametes by meiosis, followed by fusion of gametes (syngamy).

Shale A sedimentary rock formed by consolidation of clay or mud.

Sheath The tubular extracellular mucilage surrounding the cellular trichome of a filamentous prokaryote.

Shocked quartz Grains of quartz showing numerous cross-cutting dark bands (deformation lamellae) formed during the ultra-high pressure deformation associated with meteorite impact.

Signor–Lipps Effect Sampling artefact causing backward smearing in time of last appearances of taxa before a mass extinction horizon due to the incompleteness of the fossil record.

Smoking gun Clear, definitive evidence for a mass extinction kill mechanism such as the Chicxulub impact crater.

Species The fundamental category of biological classification, ranking below the genus and in some species composed of subspecies or varieties; of various definitions, the most common is the *Biological Species Concept*: 'Species are actually or potentially interbreeding natural populations which are reproductively isolated from other such groups.'

Spore The haploid product of meiosis in plants.

Spore plant Any of various 'lower land plants' that instead of producing seeds (as do gymnosperms and angiosperms) reproduce by shedding spores, such as club-mosses (lycophytes) and horse-tails (sphenophytes).

Stishovite A variety of quartz formed under very high pressure.

Stomatal index A measure of the density of stomata (gas exchange holes) on the surface of leaf that provides a valuable indication of past atmospheric carbon dioxide concentrations.

Strangelove ocean Named after the character played by Peter Sellers in Stanley Kubrick's 1962 movie *Dr Strangelove*, this refers to oceans in which all photosynthetic activity has shut down due to some cataclysm such as global darkness following a giant meteorite strike.

Stromatolite An accretionary organosedimentary structure, commonly finely layered, megascopic and calcareous, produced by the activities of mat-building microorganisms, principally filamentous photosynthetic prokaryotes such as various types of cyanobacteria.

Taxon   Any unit of the taxonomic hierarchy, from a species upwards through a genus, a family, an order, a class, a phylum to a kingdom.

Taxonomic family   In biological classification, a major category ranking above the genus and below the order.

Taxonomy   The description, naming and classification of organisms.

Tektite   Small spheres of glass formed by melting of rock at meteorite impact sites and often scattered many kilometres beyond the crater.

Toarcian   An interval of the early Jurassic, roughly 180 Ma ago, which saw a relatively minor extinction event.

Transgression   Landward movement of the coastline, as often happens during sea-level rise.

Tree of Life   A branching, tree-like representation showing the relatedness of all living organisms; commonly based on comparison of rRNAs, the ribonucleic acids of protein-manufacturing ribosomes.

Trichome   The living cellular part of a sheath-enclosed microbial filament.

Trilobite   Extinct arthropod animals of the Palaeozoic Era (543 to 245 Ma ago), characterized by a three-lobed bodily organization.

Uniformitarianism   The doctrine that the present is the key to the past, i.e. processes we can observe operating at the present day are capable of explaining all that is evident in the geological record (cf. Catastrophism).

Vertebrate   Any member of the subphylum Vertebrata that consists of all animals with a bony or cartilaginous skeleton and a well-developed brain, such as fishes, amphibians, reptiles, birds and mammals.

Vesicle   A sac-like intracellular body; in planktonic cyanobacteria and other prokaryotes, a gas-filled pocket used to control buoyancy.

Zygote   The diploid cell formed by fusion (syngamy) of two haploid gametes, the earliest formed cell of the embryo of animals or the spore-producing generation of plants.

Index

(NB: 'defn' refers to glossary entries)

Lightning Source UK Ltd.
Milton Keynes UK
24 November 2009

146651UK00002B/21/P